FLAME PHOTOMETRY: LABORATORY PRACTICE

FLAME PHOTOMETRY: LABORATORY PRACTICE

RNDr. JOSEF DVOŘÁK
RNDr. IVAN RUBEŠKA, CSc.
ZDENĚK ŘEZÁČ

English translation edited by
R. E. HESTER, B.Sc., Ph.D.
Lecturer
Department of Chemistry
University of York

ILIFFE BOOKS
LONDON

THE BUTTERWORTH GROUP

ENGLAND
Butterworth & Co (Publishers) Ltd
London: 88 Kingsway, WC2B 6AB

AUSTRALIA
Butterworth & Co (Australia) Ltd
Sydney: 20 Loftus Street
Melbourne: 343 Little Collins Street
Brisbane: 240 Queen Street

CANADA
Butterworth & Co (Canada) Ltd
Toronto: 14 Curity Avenue, 374

NEW ZEALAND
Butterworth & Co (New Zealand) Ltd
Wellington: 49/51 Balance Street
Auckland: 35 High Street

SOUTH AFRICA
Butterworth & Co (South Africa) (Pty) Ltd
Durban: 33/35 Beach Grove

Translated by Dr. O. Sofr

First published in 1971 by
Iliffe Books, an imprint of
the Butterworth Group, in co-edition with
SNTL—Publishers of Technical Literature, Prague

ISBN 592 01234 4

Printed in Czechoslovakia

CONTENTS

Preface 9

PART I **General** 11

1. History of flame photometry 11
2. Principles and characteristics of the method 15
3. Basic concepts 18
4. Different ways of using flame photometric techniques 21

PART II **Theoretical** 25

5. On the emission and absorption of radiation 25
   5.1 Basic concepts 25
   5.2 Absorption of radiation 30
   5.3 Emission of radiation 33
6. The flame 37
   6.1 Combustion 37
   6.2 The flame structure 40
   6.3 The flame temperature 45
   6.4 The flame spectrum 48
   6.5 The mechanism of excitation in flames 50
7. Inflow of the sample to the flame 51
   7.1 Possibilities of introducing the sample into the flame 51
   7.2 Dispersion 52
   7.3 The size of aerosol particles 53
   7.4 Sedimentation of aerosol particles 57
   7.5 Aerosol coagulation 57
   7.6 Evaporation in the mist chamber 58
   7.7 Evaporation in the flame 61
   7.8 The significance of theoretical observations for practical analysis 62

PART III **Practical** 64

8. Instruments and equipment 64
   8.1 Gases, equipment for their storage and purification, flow into the apparatus, and regulation 65
   8.2 Burners 72
   8.3 Dispersing devices 77
   8.4 The mist chamber 83
   8.5 Isolation of the required radiation 88
   8.6 Radiation detectors 98

9. Instruments used in flame photometry 109
   9.1 Emission analysis instruments 109
   9.2 Instruments for absorption flame photometry 117

10. The technique of work with a flame photometer 126
   10.1 Laboratory equipment 126
   10.2 Installation and adjustment of the flame photometer, illustrated with reference to the Zeiss Model III 127
   10.3 Measurement with a flame photometer 132
   10.4 Maintenance of flame photometers 137
   10.5 Preparation of solutions for analysis 138
   10.6 Determination of the analytical results 143
   10.7 Procedures of evaluation for eliminating interfering effects 148
   10.8 The technique of work in absorption flame photometry 156

11. Accuracy and sensitivity of the method 158
   11.1 Accuracy 158
   11.2 Sensitivity 162

12. Interfering factors causing erroneous results 166
   12.1 Interfering influences of transport and their elimination 168
   12.2 Interfering influences of the solid phase and their elimination 171
   12.3 Interfering influences caused by a shift of equilibrium in the flame, and their elimination 180
   12.4 Radiation interferences and their elimination 193
   12.5 Buffer solutions in flame photometry 197
   12.6 The influence of organic solvents 198

13. Emission and absorption spectra of elements 202
   13.1 General properties 202
   13.2 Spectra of the individual elements 213

14. Practical applications of flame photometry 236
   14.1 Water analysis 237

14.2 The application of flame photometry in the chemical industry 245
14.3 Analysis of silicates, minerals, ores and rocks 259
14.4 Applications of flame photometry in agricultural chemistry 269
14.5 Applications of flame photometry in biochemistry 272
14.6 Applications of flame photometry in other fields 273

APPENDIX I. Table of spectral lines and bands 275
APPENDIX II. Amounts of compounds to be weighed in preparing some frequently employed standard solutions 279
APPENDIX III. Citations of literature describing analytical methods 280

References 293

Index 317

PREFACE

Flame photometry is one of the physico-chemical methods which are becoming increasingly important in industrial, research and other laboratories. Although this method is rather young, it already has today as important a position as polarography, potentiometry, absorption photometry, etc. This is best shown by the great number of publications in specialised journals devoted to it. For example, 6 papers were published on this topic in the entire world's literature in 1945, in 1950 this number was 58 and in 1958, 213. The popularity of flame photometry is mainly due to the fact that it permits a rapid and convenient determination of alkali metals, which are difficult to determine by other means. Other methods are laborious and have low accuracy. The fact that this method is well suited to routine analysis of large numbers of samples has contributed to its wide-spread application in testing raw materials, products, water, biologic materials, etc. However, its application is not limited to alkali metals only (although this continues to be the chief application): flame photometry is increasingly used in the analysis of a number of other elements too, mainly thanks to progress in instrumentation.

The development of this method is not yet concluded by far. New possibilities of application are being tested; the sensitivity and accuracy of the method are being improved. A flame photometric technique based on a new principle, absorption flame photometry, has been developed. Technical progress has also brought a number of theoretical problems, the complete solution of which may only be expected in the future.

We believe that the evolution of this method is advanced far enough now to justify a monograph on the topic. We have tried to arrange the text in such a way as to offer the principles and procedures for practical applications of flame photometry. Chapters 8 to 15 are mainly devoted to these problems. Obviously, the theoretical part of the book (mainly Chapters 5 to 7) need not be studied for the purposes of routine analysis and practical application of the methods described in the book. It will be understood, however, that perfect application of the technique of flame photometry makes it essential to know at least the basic theoretical relationships and principles: in our opinion, these sections are discussed in the book in such a manner as to offer the basic information to analysts who wish to obtain deeper insight into theoretical problems. The purpose of the literature citations given at the respective places is to facilitate a deeper study of the topic. We wish this book to become an everyday aid to analysts in all laboratories. If it will contribute to increased application of flame photometric methods, it will fulfill its purpose.

It is our pleasant duty to thank Prof. Dr. Ing. B. Polevoj, CSc., and Ing. V. Svoboda for many valuable discussions of our work, and Ing. K. Kadič for his thorough reading of the manuscript.

J. V.
I. R.
Z. Ř.

PART I

GENERAL

1. History of flame photometry

As early as the beginning of the last century, a number of authors investigated the properties of radiation emitted by a flame into which metal compounds had been introduced. The means available at that time, however, were primitive, so that this work was only of a qualitative character. BUNSEN'S and KIRCHHOFF'S studies may be regarded as being the first in this field consciously oriented to analysis and theory. These authors not only determined the line character of the radiation from elements, but they also explained other phenomena, e.g., reversal of lines. Their results enabled some elements to be identified, so that these authors as well as their successors utilised this method in the discovery of new elements unknown up to that time. BUNSEN designed a burner for this research, which later came to be applied widely in laboratory techniques. At that time, i.e., in the second half of the 19th century, rather little attention was given to the quantitative aspects of the newly discovered phenomena, on the one hand because no suitable equipment was available, on the other due to the great sensitivity of the method, which at the time was more of a hindrance than an advantage for analytical purposes. The radiation emitted by flames was usually observed visually, or by means of rather primitive spectroscopes.

At the end of the last and the beginning of the present century, some authors, e.g., GOUY [328], BECKMANN, WAENTIG [49], KLEMPERER [444], experimented with the quantitative application of flame spectrography, based mainly on visual comparison of the radiation intensity of standards and samples. GOUY'S work is of particular significance, since this author was the first to use a concentric dispersion device to introduce sample solutions into the flame in the form of aerosols: he was likewise the first to study the relation between the radiation and composition of the solution analysed. Among other problems, GOUY discovered for example that sulphates and phosphates influence the radiation intensity of calcium and barium. These results found no major application in practical work. Nonetheless, the authors named have contributed greatly to the development of our knowledge of the properties of flames and radiations. For example, they worked out a method of measuring the temperature of flames, called the line reversal method, which continues to be used in the original way today. Quantitative application of radiation emitted by a flame only started with LUNDEGARDH [48], who is the true founder of the method. He designed an apparatus, the principle of which continues to be used in our time. LUNDEGARDH introduced a sample dissolved in water into the flame in the form of an aerósol, using a pressure dispersing device. The main part of his apparatus was a spectrograph which served to decompose light: to detect the radiation obtained he usually used a photographic plate, as is still done conventionally in emission spectrography. LUNDEGARDH was no analytical chemist in the true sense of this word: he worked in plant physiology, and he developed this method in order to be able to study the distribution of some elements in plants. The apparatus which he constructed permitted him to complete a number of studies of fundamental importance, for example in the field of plant nutrition. At the same time, however, it was found that the newly developed method was of great significance for all of analytical chemistry. LUNDEGARDH strived to perfect the original apparatus, mainly to simplify the technique of radiation detection and to increase the sensitivity of the method. The state of technology in the twentieth and thirtieth years of this century did not, however, allow him to carry out this project, and thus the designs continued to be ideas only, without being put into effect. For similar reasons, LUNDEGARDH'S successors, JANSEN, HEYES and RICHTER [411], who used a monochromator and photo-tubes instead of the spectrograph and photographic plate, did not succeed either. The equipment was too complicated, and was subject to numerous defects.

It was only in 1936 that SCHUHKNECHT [684] substantially simplified and modified the entire apparatus: it is thanks to his work that the technique of flame photometry rapidly penetrated into practical analytical chemistry. SCHUHKNECHT was the first to practically utilise the simple character of the spectrum of alkali metals and alkaline earth elements. He replaced the earlier complicated optical devices by glass filters of the same type as those which were used in colorimetry. They sufficed to isolate the required radiation, although the range of wavelengths which they transmit is relatively broad. These filters are characterised by high light transmittance, so that a simple photovoltaic cell could be used in conjuction with a suitable galvanometer. At first, this apparatus was used exclusively in agricultural chemistry. Due to its simplicity this instrument was relatively cheap. From the point of view of laboratory practice the application of this apparatus caused a revolutionary change, since quantitative analysis of alkali metals had always been a very complicated and laborious process, not to mention the low accuracy of the results. The new instrument permitted analyses to be carried out rapidly, with results which were frequently subject to errors far lower than before.

For some time, the practical utilisation of this instrument was slowed down by the 1939–45 war: soon after, however, the new method spread rapidly and continued to be perfected in many aspects. First of all, glass filters were replaced by interference filters, which were developed roughly at the same time, or by monochromators. In this way, very good isolation of the radiation studied became possible. Likewise, far more sensitive and perfect equipment was designed for radiation detection, for example photo-tubes and photomultipliers with amplifiers, which also permitted the application of flame photometry to be broadened to include other elements. From the technical point of view, possibilities of more sensitive radiation detection techniques were studied, as well as the possibilities of using higher temperatures of flames having special properties. VALLEE, BAKER, and BARTHOLOMAY [773], [774], tested the dicyan-oxygen flame; other authors use hydrogen-peroxochloride-fluoride [693], [676], or fluorine-hydrogen flames [144], [816]. A burner was also designed and used [47], which at the same time played the role of a dispersion device and permitted immediate introduction of the solution into the flame. MARGOSHES and VALLEE [500] also designed a quantometer for direct determination of the analytical results, analysing several elements at the same time. Finally, a double-beam instrument was constructed

[66] permitting some interfering influences in the analysis to be eliminated by means of using an internal standard.

From the methodological point of view, the possibilities were studied of extending the use of flame photometry to the largest possible number of elements. Much experimental material dealing with these applications was published by DEAN, et al. [170–178], [233], [524]. With, perhaps, the exception of selenium, the possibility of determining all elements was tested. Also, development was oriented to analyses of the most varied materials. Nowadays, flame photometry is being applied not only in industrial control laboratories, but in many other fields of human activity also, ranging from medical laboratories to crime investigation.

With regard to theory, various relations between the properties of solutions, flames, and sample composition were elucidated in rather a satisfactory manner. Besides SCHUHKNECHT, ALKEMADE [11–17] contributed greatly to this effort: he dealt with the mutual influence of elements, coagulation of aerosols, and the causes of varying emission intensity with changes in the solvent. Up to the present, however, not all the problems could be solved, although various authors have succeeded at least in formulating some of them correctly from the theoretical point of view, and many relatively satisfactory hypotheses have been suggested. Among others, GAYDON, et al. [298], [299], JAMES and SUGDEN [409], [410], [743], and HULDT [389], [465] investigated the properties of flame-emitted spectra. Thanks to the work of these authors, some properties of molecular spectra could be explained.

In recent years, the new method of absorption flame photometry was developed. Although the underlying principles were known for a long time, having already been studied, for example, by BUNSEN and KIRCHHOFF, it was only in 1955 that WALSH [787] suggested the use of absorption flame photometry as an analytical method. It is interesting that this procedure first found a field of application in agricultural chemistry.

The first commercial instruments for atomic absorption, as the new method was sometimes called, were soon available: likewise, the number of papers dealing with this topic has been very great in recent years. The pioneers of this method, mainly ALLAN [18–25] and DAVID [159–167], not only investigated the practical applications, but they also investigated the revelant theoretical relationships, so that in this respect progress in absorption flame photometry was very rapid.

The flame, as a medium for preparing atomic vapour, is also utilised in a method called atomic fluorescence, developed by WINEFORDNER et al. [821], [822]. This analytical procedure is not yet being used practically, as the method is too young to allow any conclusions to be made in respect of its applications. Considering the superior sensitivity for determination of some elements by means of this method, it may be assumed that it will be very useful.

2. Principles and characteristics of the method

Compounds of some elements, introduced into the non-luminous flame of a BUNSEN burner, give it a characteristic colour. This fact is well-known from qualitative analysis, being used to identify alkali metals and alkaline earths. Sodium salts colour the flame yellow, lithium salts red, etc. The colour of the flame is caused by the fact that atoms or molecules of the elements emit radiation.

Radiation may be characterised by its wavelength, which is usually expressed in nanometres or millimicrons ($1 \text{ nm} = 1 \text{ m}\mu = 10^{-9}$ m), or in angströms ($1 \text{ Å} = 10^{-10}$ m). The term wavelength will be explained in more detail in chapter 5. Here, let us only mention as an example, that light of wavelengths 400 to 435 nm is violet, 435 to 490 nm blue, 490 to 560 nm green, 560 to 595 nm yellow-green to yellow, 595 to 610 nm orange, and 610 to 770 nm red. Radiation of a single wavelength is called monochromatic. Mostly, however, atoms and molecules emit radiation of several wavelengths simultaneously. This radiation, which atoms or molecules of a certain element are capable of emitting, forms the spectrum of the atom or molecule in question. Spectra of different atoms differ from one another: the same applies to molecules. The spectrum of every atom of molecule is one of its characteristic properties. For this reason, the spectra of elements can be used to identify them.

The character of the emitted spectrum depends on temperature. As long as the radiation of flames is studied at temperatures little higher than 3200 °C, relatively simple spectra of the individual elements are observed. The human eye is capable of observing light in the wavelength range of roughly 400 nm to 760 nm. In this spectral region, in the ultraviolet (220 to 400 nm), and in the infrared (760 to 900 nm), the most typical radiation of individual atoms and molecules is found. Observation of the flame colour (into which an element or compound, capable of giving it a characteristic colour, is introduced) by the naked eye is only useful for rough orientation, since it is impossible in this way to distinguish all the radiations emitted by the flame. When, however, the radiation is observed with the aid of a spectroscope, the light being decomposed for example by a prism, the spectrum is found to be composed either of lines or of bands. Atoms give a line spectrum, while molecules give a band spectrum. Every line corresponds to a certain wave-length. The width of even this line, however, is not infinitely small: it encompasses a very small yet definite range of wavelengths, of the order of fractions of a nanometre. A band covers a larger range of wavelengths, generally in the order of ten to one hundred nanometres. Besides the types of spectra mentioned, so-called continuous spectra should be mentioned for the sake of completeness: these are mainly emitted by solid and liquid substances heated to a very high temperature. Thus the spectrum of every element consists of a greater or smaller number of lines. Where molecules are present in the flame as well as atoms, bands will appear in the spectrum as well as lines.

Another important property of every radiation is its intensity (this term will be defined in more detail in chapter 5). All lines are not equally intense in the spectrum of one element, and different elements do not emit radiation of equal intensity in a flame. Moreover, intensity depends on the flame temperature. Obviously the intensity of any individual line of a specific element depends, at a specific temperature, on the concentration of atoms of that element in the flame. When, therefore, the relation between intensity and concentration is known, it is possible to determine the concentration of a given element in the sample by means of a suitable arrangement of the inlet of the material studied into the flame.

The above principle underlies the method of emission flame photometry. The sample is usually introduced into the flame continously in the form of fine droplets of a solution, dispersed in air or oxygen with the aid of some dispersing device. In practical analysis, a flame photometer is used, the principle of which is illustrated in Fig. 2.1.

A combustible gas is introduced into the burner (H_2, C_2H_2, etc.). together with air or oxygen in which fine solution droplets are dispersed. The gas mixture at the burner outlet burns with a flame which emits the required radiation. A suitable device (a monochromator or light filter) is used to isolate that part of the spectrum of the emitted radiation which contains the line or band of the element analysed. This radiation falls onto a radiation detector (photo-tube, photo-voltaic cell, etc.) where it is converted into electrical power. The value which corresponds to the intensity of the line or band of the respective element, is read on a measuring instrument (mirror galvanometer, amplifier and milliammeter). The concentration of the element in the solution analysed is then obtained from the measured value by the use of a calibration curve (obtained from solutions of the element analysed, the concentrations of which are accurately known).

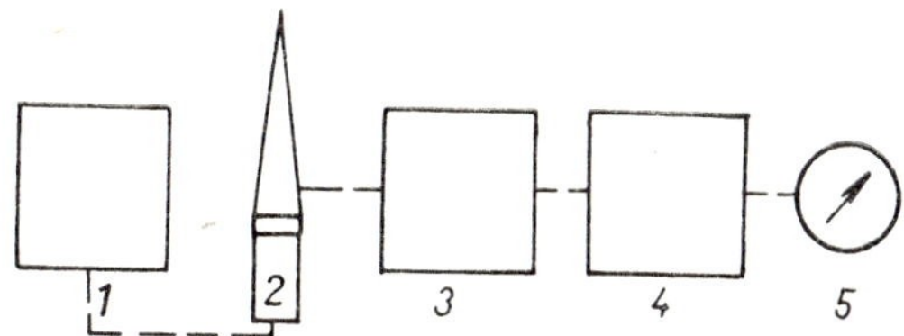

Fig. 2.1. Schematic diagram of a flame photometer
1 — inlet of gases and aerosol, 2 — burner, 3 — optical system, 4 — radiation detector, 5 — indicator

Besides the above-described method of emission flame photometry, atomic absorption photometry has been developed more recently. This is based on the well-known fact that the vapour of an element which contains free atoms absorbs light of the same wavelength as that which the atoms of this element are capable of emitting. When a light ray of this wavelength passes through a flame containing free atoms of the respective element, the intensity of the ray is diminished due to absorption. The intensity decrease thus serves as measure of the concentration of atoms in the flame, and therefore also in the solution. Differing from emission-type instruments, atomic absorption photometry instruments are fitted with a light source emitting a line spectrum of the element analysed.

The above description of emission and absorption flame photometry is greatly simplified, but in can be used in order to illustrate the degree of simplicity of the above-mentioned types of analyses. In this respect these are unequalled by any other analytical technique (with the exception of some radiometric methods).

Assessing the flame photometric technique from the point of view of practical applications in analytical chemistry, we find the following advantages:

1. a high degree of productivity of labour in routine analyses
2. in many cases, analyses can be carried out without any preceding separation procedure
3. several elements can be determined simultaneously from one solution
4. the analyses are simple to carry out, and the apparatus may be operated by staff having no special training
5. the consumption of the solution analysed is small
6. in case there are any doubts as to the accuracy of the analytical result, the measurement can be repeated with only a slight loss of time
7. the analyses are sufficiently precise
8. the method can be used to analyse some elements which can be determined by other methods only with great difficulty.

3. **Basic concepts**

Flame photometry is one of the physico-chemical methods of analysis. Its exact position in the system of physico-chemical methods is rarely mentioned in the literature, and the terminology employed frequently makes no distinction between the possible variants of this technique. Due to some similarities or relationships to other procedures, however, a range of methods must be defined which can be grouped under the name of flame photometry, and the differences between this and related techniques must be specified.

Flame photometry belongs to the group of optical methods. Emission flame photometry, like emission spectrography, studies the properties of radiation emitted by a suitable source of light. In spectral analysis an electrical discharge is used as the light source, while in flame photometry, as the name implies, a flame is used which, in general, may be obtained by combustion of any mixture of combustible gases. There is a number of similarities between the two techniques. In both cases a spectrum is studied, i.e. light emitted by the source split into its component wavelengths: the difference arises from the fact that the temperature of the flame is nearly always lower, often very much lower than that of the discharge. For this reason the spectrum obtained with the use of a flame is substantially simpler, being limited to a few of the most intense lines of the element in question. With a lower flame temperature some elements are not excited at all, so that they cannot be determined: similarly, many compounds introduced into the flame do not dissociate completely; the latter circumstance, however, causes a molecular spectrum to be obtained, which in some cases may also be employed for analytical purposes. These differences are markedly manifested in the respective instrumentation—the design of flame photometers is generally very simple.

Individual instruments designed for flame photometry differ from one another substantially, mainly in the method used to isolate light of the required wavelength and in the manner of detection of the isolated radiation. According to these factors, the methods described as *emission flame photometry* may be classified. *Flame photometric techniques* in the narrow sense include only those where radiation of a specific wavelength is isolated with the use of light filters. The group of *flame spectrophotometric methods* includes those techniques where the instruments are fitted with monochromators, and after decomposition of the light the radiation studied is isolated by means of suitable slit systems. The instruments of the latter group are more complicated. The spectrophotometer isolates a narrower wavelength interval, and therefore the procedure is more selective. A spectrograph may also be used to decompose the light of the flame, the isolated radiation being recorded on a photographic plate. In such a case we speak of *flame spectrography*. If the decomposed spectrum were studied by visual means only, the procedure would be called *flame spectroscopy*. When several elements are measured at the same time, the method involved is called *flame spectrometry*.

The second type of method is *absorption flame photometry*. As the name already implies, this is based on a different principle, i.e. studying the absorption of monochromatic light emitted by a hollow cathode discharge lamp or other suitable source. And again, depending on the means used to measure the attenuation of the original light intensity in the flame, methods may be classified as flame *spectrophotometry, spectrography, spectrometry*. This classification, however, is not used in practice. Filters are used less frequently in this method. Besides the name absorption flame photometry, the term *atomic absorption* also is frequently used.

One more method deserves to be mentioned: atomic fluorescence, in which the flame is used, as in atomic absorption, to produce atomic vapour. The energy needed to excite the atoms is supplied not by the flame, however, but by a different source of intense light.

Although the two major procedures described, i.e. emission and absorption flame photometry, are different in principle, yet they are very similar from the point of view of methods and with a full range of accessories, many instruments may be used for both purposes. Numerous relationships and influences on the character of radiation are the same with both procedures, and they are treated together in many sections of this book.

The above classification is the result of an attempt at rather an exact system of procedures, which may in principle be summed up under two headings: the one grouping emission, and the other absorption flame photometric techniques. The term *flame photometry* is sometimes used to include all the techniques discussed here, although this practice is not quite exact. However, this practice is quite acceptable for the one reason that the majority of relationships which apply to flame photometry in the more narrow sense of the word, also apply to spectrometry, spectrophotometry, etc. Methods developed with the use of an instrument which is equipped with filters, i.e. a photometer, can be used in most cases with a spectrometer or spectrograph with no modification. The reverse statement is not quite true: this, however, is due rather to the properties of the instruments involved than to the processes which take place on dispersing the sample or in the flame. Moreover, the name of flame photometry has come to be known rather widely and all related techniques are conventionally included under it. Therefore, this name is usually used to mean a method in which light, emitted or absorbed by a flame, is studied. The type of instrument to be used is then chosen according to the specific problem on hand.

4. Different ways of using flame photometric techniques

The analytical application of radiation emitted or absorbed by different atoms or molecules in a flame, is extremely varied, and therefore a number of individual types of methods may be included under the broad concept of flame photometry.

For historical reasons, let us first discuss *qualitative analysis,* which can be done either visually or with the aid of different instruments. In the so-called dry reactions of analysis, flame photometry is used for qualitative identification mainly of alkali metals, alkaline earth metals, and of several other elements, e.g. copper, boron, etc. As far as the individual radiations which are characteristic of each element can be resolved by means of suitable filters, or by decomposition of light in a monochromator, the individual elements can be identified with a high degree of reliability. With suitable instruments it is even possible to identify rubidium and caesium, which emit most intensely in the infra-red part of the spectrum, which is invisible to the human eye under normal circumstances. There are no other reliable means of identifying these two elements, with the exception of spectral analysis. Obviously, thorough knowledge of the flame – emitted spectra is a prerequisite for the identification of elements,

Mainly, however, and practically exclusively, flame photometry is used for *quantitative analysis.* It may be said without any overstatement, that it became possible to determine many elements reliably and rapidly only after flame photometric techniques were discovered and applied in laboratory practice. For example, the determination of rubidium and caesium became a practical possibility only with the advent of flame photometry. In as far as the spectrum of the element analysed is studied, and its concentration is determined by means of a calibration curve depending on the rise of the emission intensity with concentration, the simplest property is utilised, i.e. the fact that radiation intensity is proportional to the concentration of the respective element in the sample.

Such *procedures* are termed *direct* ones. Moreover, flame photometry may also be employed for indirect determinations, thus broadening the range of elements which may be analysed. *Indirect methods* are based on a chemical reaction in which, for example, the element to be analysed forms an insoluble and well-defined compound with another element which emits radiation when introduced into a flame. Such a compound may be isolated, re-dissolved, and the concentration of its inactive component may be calculated from the radiation intensity of the other component, stoichiometric relations being used as conversion factors. Such procedures are frequently used in other branches of quantitative analysis with much success (e.g. the volumetric determination of calcium): they are equally useful in flame photometry. Methods of this type have been worked out for the determination of sulphates, chromates, chlorides, etc.

It has been found that there is a number of factors in flame photometry which interfere with the determination of some elements. When direct procedures are used, these interferences must be taken into account and eliminated as far as practicable. In other cases, however, these interferences may be put to good use, enabling the determination of some elements which cannot be determined by direct means.

This involves the utilisation of some processes which take place in the flame and which interfere in the determination under normal conditions. For example, many anions decrease the radiation intensity of alkaline earth elements, forming with them thermally stable compounds which do not dissociate due to the low temperature of the flame and thus are incapable of emitting radiation. In other cases the thermal stability of these compounds will hinder the evaporation of other elements. Some components which cause quenching may be determined from the degree with which they decrease the radiation intensity of alkaline earth elements: this is sometimes called the "quenching effect". The decrease of radiation intensity is proportional to the concentration of the interfering element under suitable experimental conditions. These procedures, which are the analogue of "decoloration" procedures in colorimetry, are simple and they have been suggested for the determination of a number of elements and other components.

In some cases, knowledge of processes which take place in the flame may be utilised to determine the point of equivalence in *volumetric determinations*. Reactions on which this procedure is based may be utilised in acid-base neutralisations, precipitation, and other volumetric determinations. These are not conventional methods, but under certain conditions they may be found to be more practicable for some kinds of samples than other methods. Obviously the method will differ somewhat from the conventional procedure of volumetric determinations, although the principle remains unchanged. Examples of the application of this method are determinations of phosphates [755], potassium hydroxide [217], and calcium [232].

Direct methods of determination can also be applied to those elements or groups which do not themselves emit radiation in a flame, but which in the presence of certain oher elements form compounds which emit a very intense molecular spectrum. This procedure was suggested, for example, for the determination of the concentration of chlorides, which give a very intense molecular spectrum with copper. Naturally this procedure is more difficult, because absolutely constant experimental conditions must be maintained, as these are of great importance for the creation of molecular spectra: this applies especially to the flame temperature.

The method of absorption flame photometry has further widened the range of applications of flame photometry, namely to those elements which are difficult to excite in a flame. The analytical result is practically independent, at least within the limits set by the type of flame used, of flame temperature variations. It is true that a number of the influences mentioned in connection with emission flame photometry also apply to the absorption method, but radiation interferences are substantially less important than in the case of emission flame photometry. The apparatus again is somewhat more complicated, but on the other hand atomic absorption is far more sensitive in the case of many elements compared with emission flame photometry. Another advantage is the fact that absorption is a more unambiguous measure of the concentration of atoms, permitting therefore in many cases a study of the distribution of free atoms in the flame. Even atomic absorption, however, is not a universal method. Elements which form molecules in the flame, and are not dissociated to a sufficient degree, are outside the range even of this method. One great advantage, however, is the fact that no adjustment of the sample or elimination of accompanying elements is required, and that a very low flame temperature suffices.

It is too early to draw any definite conclusions as to the scope of application of this method, because it is a new method, which only a few years ago was described in the literature for the first time. It may be expected that this method will be perfected even more in terms of apparatus as well as of technique, a fact which also applies to all the other methods mentioned. Mainly new power sources are being sought: this does not only mean higher temperatures, but also new compositions of the combustible gas mixture, which are expected to offer the possibility of obtaining certain types of spectra to permit new special applications of the method.

It may also be asumed that the field of application of flame photometry will further widen to include a range of chemical separation methods in combination with the final flame-photometric determination. An example, selected at random, is the combination of flame-photometric techniques with ion-exchange, extraction, and similar procedures.

PART II

THEORETICAL

5. On the emission and absorption of radiation

5.1 Basic concepts

As already stated in the preceding chapter, the concentration of an element is determined by measuring the intensity of radiation in a certain wave-length range.

Radiation may be viewed as electromagnetic waves or as a flux of particles having zero rest mass, photons, which may be absorbed or emitted by other particles, e.g. atoms. Absorption transfers the energy of an electromagnetic field onto atoms; emission of radiation, on the contrary, transfers energy from atoms to the electromagnetic field. We generally characterise radiation by its wavelength, or other quantities simply related to wavelength. The relations between these quantities may be expressed by the equations

$$\lambda = \frac{1}{\tilde{\nu}} = \frac{\boldsymbol{c}}{\nu} = \frac{2\pi \boldsymbol{c}}{\omega} \tag{5.1}$$

where λ is the wavelength, cm, $\tilde{\nu}$ the wave-number, cm^{-1}, ν the frequency, s^{-1}, ω the angular frequency, s^{-1}, and lastly, $\boldsymbol{c}$ the velocity of light, $\mathrm{cm} \, . \, \mathrm{s}^{-1}$.

Considering radiation to be a flux of photons, we generally characterise them by their energy E, for which the following relation holds

$$E = \boldsymbol{h}\nu \tag{5.2}$$

where $\boldsymbol{h}$ is the Planck constant.

Thus, the radiation wavelength is decisive for the energy of a photon: the shorter the wavelength, the greater the energy of the respective photon. The radiation generally used in flame photometry ranges from the ultraviolet region (λ = 200 nm) to the near infrared (λ = 900 nm), the majority of measurements being, of course, carried out in the visible light region (400 to 750 nm).

All substances are capable of absorbing and emitting electromagnetic radiation in the solid, liquid and gaseous state. The dependence of emission or absorption on wavelength is called, as already mentioned, the spectrum of the respective substance. The wavelengths of the emission and absorption lines of the spectrum are identical, except when fluorescence phenomena are involved. Atomic spectra are of prime importance in flame photometry, and therefore we shall discuss them first.

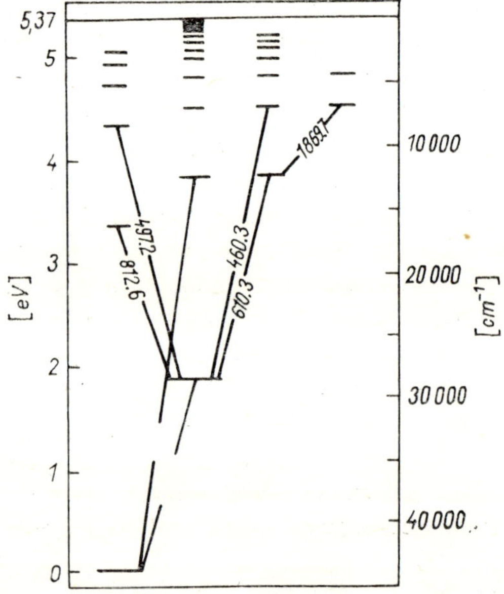

Fig. 5.1. GROTRIAN *diagram of lithium terms*

According to the quantum theory, atoms may exist in certain *energy levels,* i.e. have certain discrete energy contents. On changing from a higher energy level to a lower one, i.e. from a state richer in energy to one of lower energy, the excess energy is lost in the form of a photon. On the other hand, an atom in the lower energy level is able to absorb a photon whose energy corresponds to the difference of energy needed to achieve one of the higher levels.

The possible energy levels are usually illustrated by means of the so-called Grotrian diagrams (Fig. 5.1). Horizontal lines denote the possible energy states of the atom, and permitted transitions are illustrated by connecting lines bearing the wavelength of the respective radiation. Each such wavelength occurs in the spectrum as a *spectral line*. The energies of the states are plotted on the ordinate in eV or in wave-numbers, $\bar{\nu}$, corresponding to these values. The relation between these two scales is

$$1\ \text{eV} \sim 8068\ \text{cm}^{-1} \tag{5.3}$$

We ascribe a zero energy value to the lowest state, calling it the ground state. The energy needed for transition of an atom to a given higher state is called the *excitation energy*. Lines corresponding to transitions back to the ground state are called *resonance lines*. In these transitions the whole excitation energy passes to the photon, so that the frequencies of these lines are given by relation (5.2). Resonance lines are the most intense lines of the spectrum. They are important for atomic absorption, since they represent radiation which the atom is capable of absorbing in the ground state. By accepting increasing amounts of energy the atom is excited to energetically higher and higher states, until at the value of the so-called *ionisation energy* an electron is totally removed from the sphere of influence of the atom. Thus the atom obtains a positive charge, becoming an ion, which is then able to form its own ionic spectrum, quite different from the spectrum of the atom. The ionic spectrum of an ionised atom is similar in structure to the spectrum of the element whose atomic number is one unit lower. Atomic spectra occur generally in the flame, ionic spectra appearing only in the case of alkaline earth elements, where they are sometimes used analytically.

Of course, even the lines corresponding to free atoms or ions are not perfectly monochromatic. They fill a finite, although very small interval of wavelengths. The dependence of intensity of a certain line on wavelength is called the *line profile*. In a flame, due to the disordered thermal motion of atoms, the profiles most frequently encountered are the *Gaussian profile*, due to the Doppler effect, and the *resonance profile*, due to collision dampening, i.e. collisions of radiating atoms with other particles. An important property of the line is its width, as will be explained later, expressed as the width at one-half the height of the line peak (see Fig. 5.2).

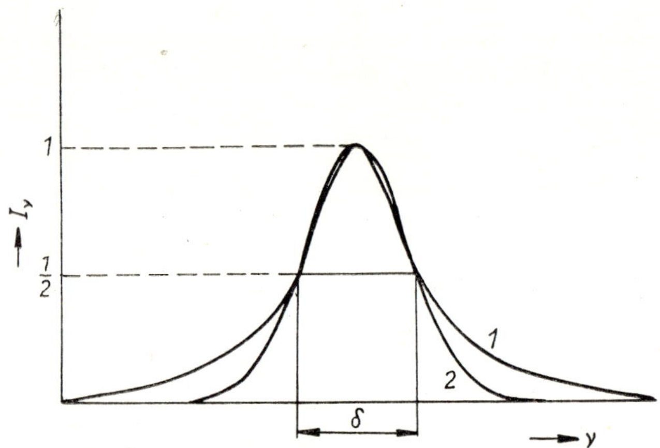

Fig. 5.2. Shape of a line of resonance profile (1) and Gaussian profile (2) of equal width δ

In emission measurements the line profile and width are usually unimportant, as the spectral interval isolated by conventional monochromators is incomparably larger than the line width. But, as will be seen from the following discussion, the line profile and width have a very great influence in absorption flame photometry.

While the energy of the free atom is given only by the energy of its electron orbits, the energy of particles with chemical bonds, i.e. of molecules and radicals, depends on their vibrational and rotational states. These energy values differ significantly in the magnitude of their energy quanta. While the electronic energy quanta are of the order of eV, those of vibrational energy are of the order of tenths of eV, and those of rotational energy are only hundredths of eV. In the transition of a molecule from one energy state to another, changes of the electronic energy are accompanied by vibrational and rotational energy changes, so that instead of a single line we obtain a whole set of lines, called a band, which corresponds to various rotational and vibrational energy changes. The character of a molecular spectrum with good resolution of the individual lines is shown in Fig. 5.3.

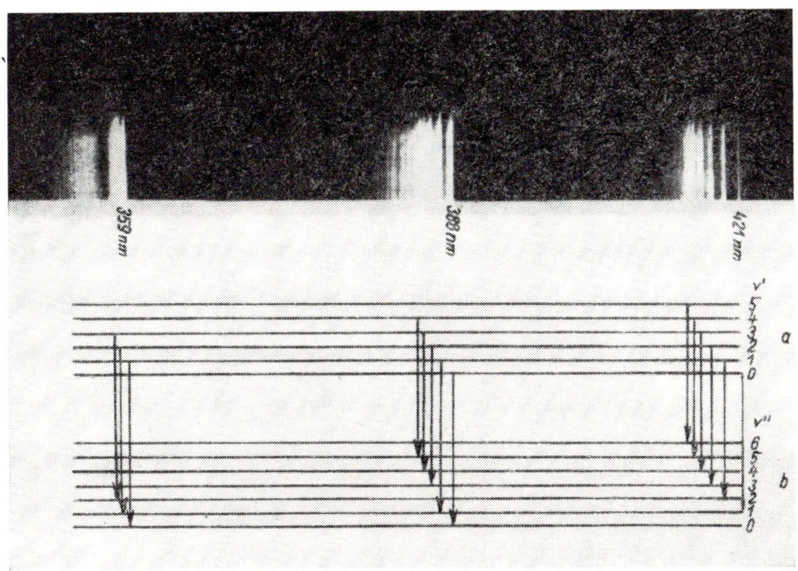

Fig. 5.3. Molecular spectrum of CN—cyano bands
a, b — upper and lower electronic states with a number of vibrational states v′ and v″. Arrows indicate transitions corresponding to the individual band edges. The rotation levels corresponding to every vibrational state are not indicated.

Before starting to discuss the basic concepts which hold for radiation absorption and emission of a flame, we must define some terms which we shall use.

Radiation flux is the total radiation energy passing through a definite area in a unit of time.

The radiation flux density, I_ν, is the radiant energy of a frequency ν, which passes in a unit of time through a unit area located in a position perpendicular to the direction of the rays, in a unit frequency range. The radiation flux density of the total radiation contained in a certain spectral line is called the *integral*. Between the two, the following relations holds

$$I = \int_0^\infty I_\nu \, d\nu,$$

the integration being carried out over the whole line profile (Fig 5.2).

The integral emission coefficient of a line, ε, is the energy irradiated by the line out of a unit volume into a unit solid angle in a unit of time.

Relating the integral emission coefficient to a unit frequency interval we obtain the *emission coefficient*, ε_ν. A similar relation holds for these two quantities

$$\varepsilon = \int_0^\infty \varepsilon_\nu \, d\nu.$$

Since a flame fulfills, to a large degree, the notion of the so-called homogenous radiant layer, i.e. a layer of constant composition and temperature in the direction of observation, the individual relations are simple. We shall concentrate our attention here mainly on line spectra, although similar relations also hold for other types of spectra.

5.2 Absorption of radiation

The well-known *Lambert – Beer* law holds for the absorption of a parallel beam of rays

$$I_\nu = I_{0\nu} \, e^{-K_\nu l}, \qquad (5.4)$$

where $I_{0\nu}$ is the density of a radiation flux incident on the absorbing medium, I_ν is the radiant flux density after passage through the absorbing layer, K_ν is the absorption coefficient, and l is the path length of the ray in the absorbing medium. The subscripts ν mean that these quantities relate to a certain frequency. The dependence of the absorption coefficient of a resonance line on the concentration of atoms may be expressed by the relation

$$\int_0^\infty K_\nu \, d\nu = \frac{\pi e^2}{mc} N f = \frac{h\nu}{c} B_{mn} N = \frac{\lambda^2 g_m}{8\pi g_n} A_{nm} N, \qquad (5.5)$$

where (as already follows from the fact that the entire left-hand side of the equation is an integral) we are dealing with the so-called *integral absorption coefficient*. N is the number of atoms in the ground state in a unit volume, f is the strength of the oscillator*, e the elementary charge, m the electron mass, B_{mn} the EINSTEIN probability of absorption, A_{nm} the probability of spontaneous emission, and g_n, g_m the statistical weights of the upper and lower energy states, respectively.

In practical measurement, we generally have no monochromatic radiation available. The measured spectral interval usually is determined in practice either by the parameters of the dispersing system, when a light source with a continuous spectrum is used, or by the line width, when a source emitting a line spectrum is used. To elucidate this state, the basic absorption equation would have to be integrated over the entire width of the spectral interval actually measured. In atomic absorption spectroscopy, therefore, light sources emitting very narrow lines, of the order of 0.01 Å, are used.

With a constant value of the integral absorption coefficient, which is related to the concentration of atoms according to equation (5.5), the value of the absorption coefficient in the line peak, K_{max}, depends on the line profile. For a line with Gaussian profile the absorption coefficient at the peak is given by the equation

$$K_{\max} = \frac{2}{\delta_D}\sqrt{\frac{\ln 2}{\pi}}\int_0^\infty K_\nu \, d\nu = \frac{2}{\delta_D}\sqrt{\frac{\ln 2}{\pi}}\,\frac{\pi e^2}{mc}Nf \qquad (5.6)$$

where δ_D is the line width, whose temperature dependence is given by the relation

$$\delta_D = 7.18 \cdot 10^{-7}\nu\sqrt{\frac{T}{M}} \qquad (5.7)$$

* The oscillator strength is a quantity which characterises the ease of a transition between two energy levels. As also follows from equation (5.5), it is proportional to the transition probability, and thus also determines the intensity of the respective line. More detailed information will be found, for example, in the book: FRISH, S. E.: Opticheskie spektry atomov (Optical spectra of atoms) Gos. Izd. Fiz. Mat. Moscow 1963.

where M is the atomic mass of the element, and T the absolute temperature.

If collision dampening exerts an interfering influence on the surrounding particles, then K_{max} is given by the equation

$$K_{max} = \frac{e^2}{\delta mc} fN \tag{5.8}$$

where δ is again the line width. Substituting relations from equations (5.6) and (5.8) into equation (5.4), we obtain, after converting to logarithms

$$A = \log \frac{I_0}{I} = \frac{2}{\delta_D} \sqrt{\frac{\ln 2}{\pi}} \frac{\pi e^2}{mc} Nfl \tag{5.9}$$

or

$$A = \frac{e^2}{\delta mc} fNl \tag{5.10}$$

where A is the absorbance or optical density. In principle, these are the equations for the absorption curves for the case where the emitted line width is negligible compared to the width of the absorbing line.

When absorbance is plotted against the concentration of atoms, the calibration curves are linear, passing through the origin. Their slopes are proportional to general constants as well as to the thickness of the absorbing layer and oscillator strength for the given line, and indirectly proportional to the line width.

In reality the plots are generally more or less curved. This deviation may be caused by the fact that the width of the radiated line is not negligible compared to the width of the absorbing line. Thus we cannot neglect integration over the width of the measured spectral interval and approximate the absorption coefficient by its value at the peak. Moreover, with increasing concentration of the element in the flame the line may widen, due to resonance dampening, leading to a lower value of the absorption coefficient in the peak, as follows from equation (5.10).

Imperfections of the experimental arrangement of the measuring instrument may also be the cause of curvature in the plots. If, for example, the entire radiation incident on the detector is not absorbed, the measured absorbance tends, with increasing concentration, to the value

$$A = \log \frac{I_0 + i}{i} \tag{5.11}$$

where i is the radiation incident on the detector which is not absorbed.

The character of the curves may also be influenced by the method of radiation detection. If the entire radiation incident on the detector is measured, curvature may appear due to the emission of the element in the flame, since this radiation is superposed over the radiation of the light source, increasing the signal measured. This influence is especially significant when lines are measured in the visible spectral region. Such lines are radiated with sufficient intensity even at low flame temperatures. For this reason the radiation of the discharge lamps is usually modulated (interrupted) and only the alternating component of the detector signal is measured. Thus the unmodulated radiation of the flame is eliminated. The methods of detection are described in section 8.6.

5.3 Emission of radiation

Free atoms in a flame are partly excited, and radiate light quanta. Although the mechanism of excitation may differ in individual elements, it still follows from experience that at a sufficient distance from the reaction zone of the flame atoms are practically in thermal equilibrium, so that their distribution into excitation states may be expressed by BOLTZMANN'S law

$$N_m = N_0 \frac{g_m}{g_0} e^{-\frac{E_m}{kT}}, \tag{5.12}$$

where N_m is the number of atoms in the state m, N_0 the number of atoms in the ground state, g_m, g_0 the statistic weights of the state m and of the ground state, respectively, E_m the excitation energy of the state m, k the Boltzmann constant, and T the absolute temperature.

The BOLTZMANN law holds in this simple form only when the number of excited atoms is small compared to the total number of atoms. It is justified to a substantial extent at the temperatures conventionally achieved in flames.

The energy radiated by a certain line from a unit volume of the flame into a unit solid angle is given by a relation similar to equation (5.2):

$$\int_0^\infty \varepsilon_\nu \, \mathrm{d}_\nu = \frac{1}{4\pi} A_{nm} N_m h\nu, \tag{5.13}$$

where A_{nm} is the probability of spontaneous emission according to EINSTEIN, and N_m is the concentration of excited atoms from equation (5.12).

A part of this energy is, of course, re-absorbed in the flame, as follows from the preceding discussion. Let us limit our discussion to the radiation of a single frequency ν and neglect forced emission, which is permissible in flame photometry. Moreover, let us consider a radiating layer of thickness dl, but of an area compared to which the area of the unit volume is negligibly small. We may now express the density of the radiation flux, radiated by this layer in a perpendicular direction, as

$$\mathrm{d}I_\nu = \varepsilon_\nu \, \mathrm{d}l, \tag{5.14}$$

At the same time, however, there is self-absorption, decreasing the above value by

$$-\mathrm{d}I_\nu = K_\nu I_\nu \, \mathrm{d}l \tag{5.15}$$

Adding the two equations, we obtain

$$\frac{\mathrm{d}I_\nu}{\mathrm{d}l} = \varepsilon_\nu - K_\nu I_\nu, \tag{5.16}$$

Since the KIRCHHOFF law holds under thermal equilibrium conditions,

$$\frac{\varepsilon_\nu}{K_\nu} = B_\nu \tag{5.17}$$

where B_ν is the radiant flux density of an absolutely black body at a frequency ν given by PLANCK's law, we obtain, after substitution into equation (5.16):

$$\frac{\mathrm{d}I_\nu}{\mathrm{d}l} = K_\nu (B_\nu - I_\nu). \tag{5.18}$$

For a homogeneous layer the quantities I_ν, K_ν and B_ν are independent of l.

The differential equation (5.18) may be solved by separation of the variables: substituting the limiting conditions ($l = 0$ and $I_\nu = 0$), we obtain the following expression for the dependence of the radiation flux density on the thickness of the radiating layer

$$I_\nu = B_\nu(1 - e^{-K_\nu l}). \tag{5.19}$$

The exponential term may be expanded into a series

$$e^{-K_\nu l} = 1 - \frac{K_\nu l}{1!} + \frac{K_\nu l^2}{2!} - \ldots \tag{5.20}$$

and for $K_\nu\, l \ll 1$ we may limit ourselves to the first two terms. Equation (5.19) then converts into

$$I_\nu = B_\nu K_\nu l, \tag{5.21}$$

indicating the direct proportionality between the radiant flux density and radiating layer thickness l, or the absorption coefficient K_ν, which is proportional to the concentration of atoms [see equation (5.5)]. This linear dependence is, of course, only valid for low concentrations. It follows from equation (5.19) that I_ν may achieve a maximum value

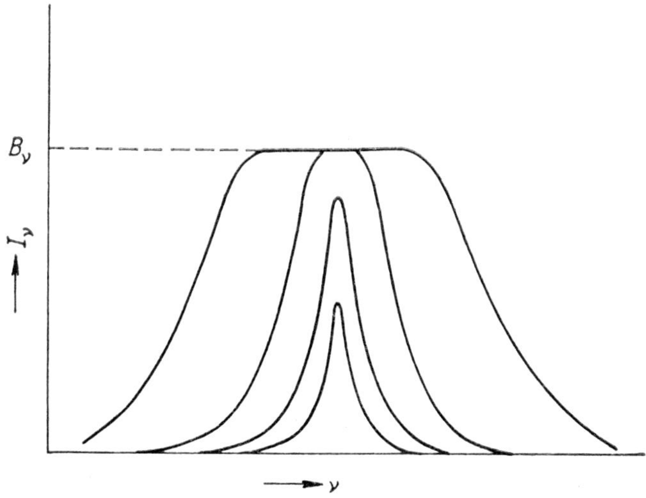

Fig. 5.4. Shape of a line with increasing concentration of the element in the flame and with self-absorption
B_ν — radiation flux density of absolutely black body

of B_ν, corresponding to the absolutely black body at the given wavelength and temperature. Thus, I_ν in the line peak rises with the increasing element concentration in the flame only up to the value of B_ν. After that only the line width increases (see Fig. 5.4).

In flame photometry, however, radiation is not usually measured in the line peak, the integrated value being used instead, since the spectral interval selected by the monochromator or filter is far greater than the line width itself. The following holds for the integral radiation flux density, according to equation (5.19)

$$\int_0^\infty I_\nu \, d\nu = B_\nu \int_0^\infty (1 - e^{-K\nu l}) \, d\nu. \tag{5.22}$$

In this equation B_ν stands before the integration sign, since it is regarded as constant within the width of the spectral line. The dependence of the integral radiation of the line on the product $K_\nu l$, which is sometimes denoted *optical density*, in fully logarithmic coordinates, is called the growth curve. Its form depends, as is evident from Fig. 5.5, on the line profile. For the resonance profile the logarithm of this quantity rises first with a slope of 1, then it bends and continues to rise with a slope of 1/2. For the Gaussian profile it only approaches a certain value and rises no more. For lines whose profiles are given by a combination of the resonance and Gaussian profiles (so-called Voigt profile), a greater or smaller delay is observed on the curve.

In flame photometry the quantity measured is generally described as *line intensity*. By plotting this measured intensity against the logarithm of the element concentration we obtain a dependence approaching the case of the resonance profile. If, however, we plot intensity against concentration in the conventional manner, we do not obtain a second linear branch, but a curve according to the relation

$$I = \text{const} \cdot N^{1/2} \tag{5.23}$$

The shapes of calibration curves are frequently expressed mathematically by means of an empirical relation, usually called the SCHEIBE–LOMAKIN equation

$$I = ac^b, \tag{5.24}$$

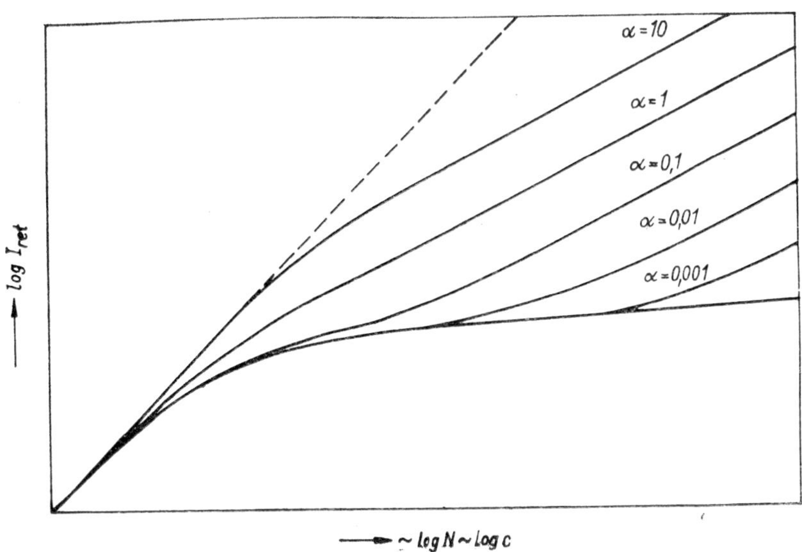

Fig. 5.5. Growth curves

Relationship between integral radiation flux density I and the element concentration N for different line profiles

$$\alpha = \delta/\delta_D$$

where c is the concentration of the element in the solution, and a and b are constants. The exponent b may achieve values of 1 to 1/2, due to self-absorption, as already stated. However, other factors also can influence the value of this exponent, e.g. imperfect evaporation of solid particles in the flame, or ionisation of the element determined. These will be discussed in chapter 12.

6. The flame

6.1 Combustion

The flame is a source of energy which is used for analytical purposes in emission and absorption flame photometry. Although it was used for a long time as a source of light as well as an important source of energy for technological purposes, our knowledge of the processes which take place in flames is still unsatisfactory, and some relationships and effects have not been completely explained up to now.

Combustion may be defined as a chemical reaction of two substances (one of which is called the fuel and the other the oxidant), taking place with intensive liberation of heat. In practice, the oxidant is understood to mean a substance which either contains atomic oxygen, or is able to liberate it in some way. In flame photometry the oxidant usually is oxygen or air, which has a number of advantages notwithstanding its relatively low percentage of oxygen.

In general, fuels may be understood to include any substance which on oxidation is capable of burning in the air. Fuels used in flame photometry are mainly gases, especially acetylene or hydrogen: other fuels are of course also used, e.g. methane, propane, cyanogen, etc. These gasous fuels permit combustion of so-called homogeneous mixtures.

Table 6.1. LIMIT OF INFLAMMABILITY AND IGNITION TEMPERATURE OF GASES MIXED WITH AIR

Gas	Lower inflammability limit, % by vol.	Upper inflammability limit, % by vol.	Ignition temperature °C
Hydrogen	4.00	74.20	530
Methane	5.00	15.00	645
Propane	2.37	9.50	510
Butane	1.86	8.41	490
Acetylene	2.50	80.00	350
Cyanogen	7.60	38.00	—
Town gas	9.80	24.80	560

Combustion propagates in a gaseous mixture only when the ratio of fuel to air or oxygen achieves a certain value called the inflammability limit. This value differs according to the experimental conditions, e.g. initial pressure, temperature, moisture content and shape of the vessel in which the combustion process takes place, direction of combustion, etc. Some data on those fuels used most frequently in flame photometry are summarised in Table 6.1 [448, 510, 511]. In order that combustion should take place at all, it obviously is essential that a certain temperature, called ignition temperature, be reached in some place. The ignition temperature values differ with different substances, depending also in many respects on the practical conditions. For the fuels which are used most often in flame photometry these values are listed in Table 6.1. Combustion of gaseous mixtures is one of the most complicated chemical reaction types. In general this is assumed to be a chain reaction. Let us take the example of hydrogen combustion [30]

1. $H_2 + M = 2\,H + M$ – primary hydrogen activation
2. $H + O_2 = OH + O$ – slow reaction, $E_2 \sim -18$ kcal/mole
3. $O + H_2 = OH + H$ – slow reaction, $E_3 \sim -6$ kcal/mole
4. $OH + H_2 = H_2O + H$ – rapid reaction, $E_4 \sim 10$ kcal/mole

E denotes the energy liberated in the respective process. The molecule M does not take immediate part in the first reaction step, but plays an auxiliary role in starting the chain by primary activation of the hydrogen molecule. The kind of molecule actually involved cannot yet be decided upon unambiguously. The second step, which is the slowest one, determines the overall rate of the chain reaction. The basic elements of the chain are hydrogen atoms, which are regenerated in the process. The entire cycle may be summarised in a reaction which is obtained by addition of the reactions 2,3 and twice 4.

$$H + 3\,H_2 + O_2 = 3\,H + 2\,H_2O$$

Thus one hydrogen atom causes three new hydrogen atoms to be formed, these in turn permit another nine hydrogen atoms to be obtained, etc. The chain branches out progressively throughout the reaction space. Hydrogen atoms are, however, also consumed in another way, namely in the reaction

$$2\,H + M = H_2 + M$$

i.e. recombination to molecular hydrogen. This reaction is strongly exothermic and its significance rises until equilibrium with reaction number 2 of the overall scheme is achieved. Since the presence of OH radicals and of atomic hydrogen in the reaction space was actually proved, and it was found that their concentration is substantially larger than under equilibrium conditions, these radicals and perhaps even atomic hydrogen must be taken to be basic active centres of this chain reaction. According to some authors, however, the participation of a hypothetical molecule HO_2 as active centre is possible, as well as that of hydrogen peroxide molecules. Similar schemes may be constructed for the combustion of carbon monoxide, which, however, from the flame photometry point of view is uninteresting. Combustion of hydrocarbons, which are more important in this respect, is far more complicated and not yet understood in sufficient detail.

6.2 The flame structure

The above-described combustion process is related, as mentioned already, to emission of heat and light. The limited space in which this process takes place is called a flame. Owing to the rates of chemical reactions, diffusion flames must be distinguished from flames in which a homogeneous pre-mixed mixture of the fuel and oxidant burns. This latter type of combustion is sometimes called kinetic combustion, the respective flame being called explosive by some authors. In diffusion combustion the oxidant and fuel are not pre-mixed, but penetrate into each other by means of mutual diffusion, and the rate of the combustion process is mainly determined by the rate of this diffusion process. Since the rate of the reaction proper is substantially greater than that of the diffusion process, it has no substantial influence on the rate of the overall process. Kinetic combustion takes place in those cases where the rate-determining process, i.e. the slowest one, is combustion proper. This situation exists when the oxidant and fuel are perfectly mixed and diffusion is insignificant, at least in the sense of the diffusion flame. In flame photometry only so-called stationary flames are of any significance. This group includes both types mentioned.

Equipment which permits a stationary flame to form is called a burner. When a gaseous mixture containing the appropriate concentrations of the fuel and oxidant is ignited in some place, a narrow chemical reaction zone is first formed, spreading from one gas layer to another by means of transmission of heat and active centres. This zone of rapid chemical reactions and great temperature and concentration gradients represents the so-called flame front, which propagates by conduction of heat from the combustion products to the combustible mixture as well as by diffusion of various compounds or radicals which are contained in the reacting mixture. The molecules of the intermediate reaction products may be active combustion centres, and therefore the rate of their motion may have a decisive influence on the rate of flame propagation, which is sometimes also called rate of combustion. This latter rate depends on many factors, e.g. the rate of flow of the combustible mixture, its temperature, pressure, composition, etc. The burner is that piece of equipment which allows the rate of combustion and rate of inflow of the combustible mixture to approach one another to give a state of equilibrium, permitting a stationary flame to be formed. In this case, thermal equilibrium is also obtained between the heat absorbed by the combustion products and the heat transmitted to the combustible mixture due to its thermal conductivity. The flame velocity is usually defined by means of its normal velocity, i.e. the velocity with which the flame front propagates in the combustible mixture in a direction perpendicular to the surface of the flame front. The combustion zone proper is preceded by a pre-heating zone. Here, the temperature rises in the direction of the gas flow in roughly an exponential manner [474]. From one side, this zone is bounded by the ignition temperature, from the other side (according to convention) by a temperature 1 °C higher than the original temperature of the mixture. The width of the pre-heating zone depends only on the pressure and rate of combustion, decreasing with the rising rate of combustion. Under normal conditions this width is only a few tenths of a millimetre, being even less with some mixtures [298].

The pre-heating zone is followed by the reaction zone, which is well visible in most flames because it emits intense radiation. Temperatures in the reaction zone rise up to a certain value, after which they remain constant in this zone. The width of this zone is small, depending on the composition of the mixture as well as on the flame velocity, being again of the order of tenths to hundredths of a millimetre, thousandths in the case of hydrogen. When the fuel and the oxidant are pre-mixed, kinetic combustion sets in in this zone, as already mentioned. The above phenomena are best explained by means of the example of the BUNSEN burner (Fig. 6.1), where the individual parts of the flame are easily recognised. With respect to the laminary gas flow the reaction zone is conical in shape (it is sometimes also called the primary zone). The normal velocity is determined by the so-called cosine or MICHELSON law [448, 824]

$$u = v \cdot \cos \alpha, \tag{6.1}$$

where u is the velocity of flame propagation at the given point,

v – the flow-rate of the gas and

α – the angle between the burner axis and the normal at the combustion surface.

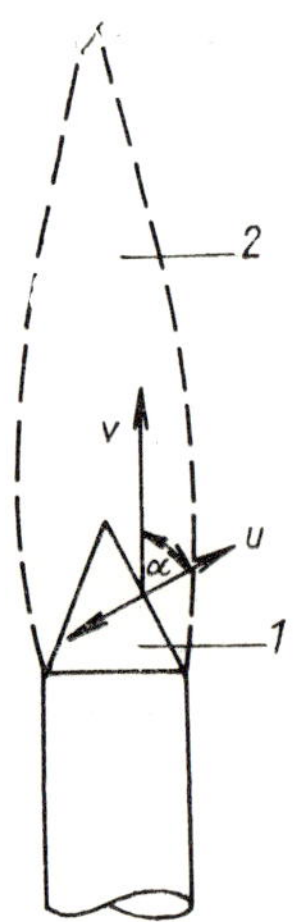

Fig. 6.1. Laminar flame

1 — primary zone, 2 — outer cone, v — rate of gas flow, u — rate of propagation of the flame

This equation indicates several relationships which are frequently utilised in the practical application of organic solvents in flame photometry. The shape of the inner cone depends on the rate of combustion, among other factors. With an increased inflow of gases into the burner, the length of the cone will increase, since the combustion of a larger amount of gas within a unit of time will likewise require a large combustion surface. With a constant load on the burner, i.e. constant flow-rate of the gas mixture, the height and therefore also the surface of the cone will decrease with the increasing combustibility of the gas mixture. Whether or not the mixture itself contains an excess of the oxidant determines the size of the flame cone. Such changes take place when a solution of the sample in some organic solvent is used instead of an aqueous solution.

Table 6.2 lists some values obtained on combustion of an acetylene-air flame with a certain air excess in a MECKER type burner belonging to the auxiliary equipment of a Model III Zeiss flame photometer. It is seen from this table that the overall height of the flame varies relatively little.

Table 6.2. DEPENDENCE OF THE FLAME HEIGHT IN AN ACETYLENE-AIR FLAME ON THE SOLVENT USED

Solvent	Cone height (mm)	
	inner	outer
Water	2.0	165
Methyl alcohol	2.9	185
Ethyl alcohol	3.8	190
n-Propyl alcohol	4.8	195
n-Butyl alcohol	6.0	200

Above the reaction zone there is a so-called external cone or secondary zone which differs from the inner cone in that it usually is quite colourless. At normal pressure it is substantially larger and may take different shapes. With the MECKER burner belonging to the Zeiss flame photometer this zone is conical. As mentioned already, diffusion-tyoe combustion takes place in this zone. This is due to the fact that either the combustion reaction proceeded only to the first step in the reaction zone [510], as is usually stated in the case of acetylene, or the reaction was not completed, as stated for example by KHITRIN [431] in the case of carbon monoxide. Thus the combustion process is completed in this external zone, the reaction rate being determined, as already mentioned, by the slowest process, i.e. diffusion.

Depending on the shape of the flame, another classification may also be used. This is rather more practical in character, being therefore specially important in flame photometry. In this case, laminar and turbulent flames are distinguished. The laminar flame only occurs with a laminar flow of the fuel and oxidant, the second case occurs with laminar as well as turbulent inflow of the combustible mixture. In the case of turbulent inflow and combustion, entire aggregates of molecules are transported, diffusion of entire groups taking place. Laminar and turbulent flow may be distinguished by the REYNOLDS number

$$\mathrm{Re} = \frac{sd\varrho}{\eta}, \tag{6.2}$$

where s is the mean flow rate,
d – the diameter of the tube or flame,
ϱ – the gas density and
η – the gas viscosity.

As long as the value of Re is less than 2300, flow is laminar: with values over 3200, turbulent flow is involved. A turbulent flame is characterised by a roaring noise, and its shape differs from that of a laminar one, as seen in Fig. 6.2. The main difference is in the inner blue cone,

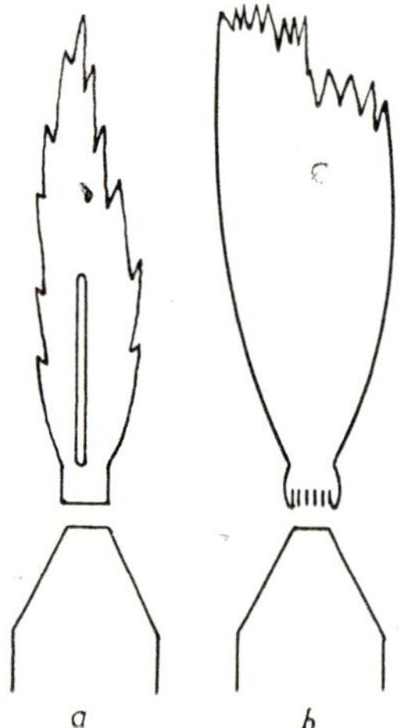

Fig. 6.2. Turbulent flame
a — simple flame,
b – in the case of dispersion of water in the flame

which is elongated in shape and stops being sharply contoured. Similarly, the shape of the outer cone changes: its contours are less distinct. The surface area of a turbulent flame is relatively larger than that of a laminar one, and therefore the overall diffusion value and rate of combustion increase (as far as one can speak of a rate of combustion in turbulent flames in the same sense as with laminar ones: at least, however, this applies to the external cone). The rate of combustion in the inner zone has not yet been studied in sufficient detail. With this flame type the mean flow rate should be controlled, i.e. it should depend on the turbulent heat conductivity, turbulent diffusion, and the chemical processes in the combustion zone.

Both types of flames can be used to study emission-spectral processes, while laminar flames are mainly used in absorption photometry, although a few cases are known in which a turbulent flame was employed. Both, however, have their advantages and disadvantages. The laminar flame is usually used in conjunction with indirect spraying of the solution studied (apparatus with mist chamber): it will be seen in the following chapters that this arrangement leads to a number of processes which, when summed together, may cause the experimental situation to become rather complicated. The turbulent flame is usually used with direct inlet of the aerosol sample into the flame, eliminating a number of complications, but involving on the other hand a number of other disadvantages. First of all the turbulent flame is inhomogeneous in itself and its volume is variable. The range of sizes of particles entering the flame is great, so that larger droplets pass through the flame without evaporating. When organic solvents are used, changes may occur which are not controllable in all cases.

6.3 The flame temperature

The temperature of a flame is a very important parameter, since it determines the proportion of excited atoms from the overall number of atoms entering the flame, and thus the intensity of the emitted radiation. The flame temperature is in itself rather a complicated phenomenon. Some authors [298] believe that it is impossible to speak of a flame temperature, and that what is measured is only the temperature of the element which is introduced into the flame for measurement by the line reversal method.

The term temperature is actually rather imprecise, if applied to a combustion zone where there is no state of equilibrium, and where therefore only an effective temperature can be determined [298, 547]. The gases in the outer flame cone are practically in equilibrium, so that the temperature is quite well defined in this region. However, the values measured differ from those calculated according to the KIRCHHOFF theorem, mainly because combustion is not a strictly adiabatic process, and because at temperatures over 2000 °C the molecules of combustion products dissociate to a substantial extent, and the energy consumed in this process must also be considered.

The line reversal method is the most advantageous of the methods used to measure flame temperatures. It is based on comparison of the flame temperature measured with the temperature of a known source of light, Na or Li being introduced into the flame to serve as comparison elements. It is assumed that these are in equilibrium with the other molecules and atoms occurring in the flame. More details of this method or of other procedures for temperature measurement cannot be presented within the scope of this book, and the reader is referred to the original literature [298, 547].

Table 6.3. FLAME TEMPERATURES

Composition of combustion mixture	Flame temperature, °C
Acetylene – air	2125
Acetylene – oxygen	3100
Hydrogen – air	2045
Hydrogen – oxygen	2660
Propane – air	1935
Town gas – air	1845
Butane – air	1895
Cyanogen – oxygen	4467

The flame temperature is determined by several factors, mainly depending on the kind of fuel and oxidant used. For the flames conventionally used in flame photometry, temperature values are given in Table 6.3 [177, 213, 373, 510, 815, 829]. Obviously, higher flame temperatures are obtained with oxygen used as oxidant instead of air, since inert nitrogen in the air lowers the flame temperature.

The flame temperature also varies according to the ratio of the individual components in the combustible mixture. If the mixture does not enter the burner in the optimum composition, the excess fuel or oxidant is ballast, i.e. does not participate in the reaction and like an inert gas this excess component lowers the flame temperature. The dependence of flame temperature on composition is shown in Fig. 6.3 [714, 216] for acetylene–air flame. The composition generally chosen is such that the emission intensity of the element to be determined is greatest. It should be mentioned here that the mixtures involved are not strictly stoichiometric, nor must it necessarily be that mixture which gives the maximum temperature. The optimum fuel-to-oxidant ratio differs with different elements, since chemical reactions in the flame may also participate in the process.

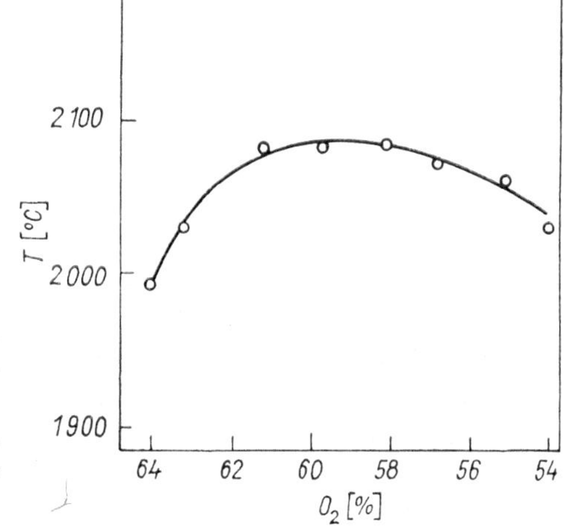

Fig. 6.3. Relationship between the temperature of the acetylene-air flame and composition of fuel gas mixture.
The molar fraction of air in the acetylene-air mixture is expressed in percentages of oxygen.

The flame temperature also is influenced by the type of solvent introduced into the flame. Irrespective of whether it is combustible or not, the evaporated solvent always is ballast in a flame burning under optimum conditions, and lowers the flame temperature.

With instruments having a mist chamber this influence is manifested to a relatively slight degree. Various authors [216], [478], [419] mention a temperature decrease in the case of water of only 20 to 30 °C. This is easily understood, considering that in a unit of time, only a very slight amount of the liquid enters the flame.

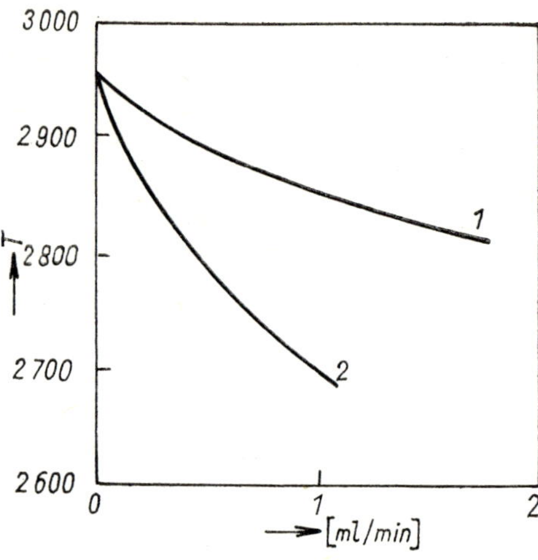

Fig. 6.4. The influence of solvents on the flame temperature in the case of a direct-injection burner. 1 — methanol, 2 — water

With a burner fitted with direct inflow of the liquid into the flame, the changes involved are substantial, since the entire amount of the solution analysed enters the flame. Obviously, substantial differences between organic solvents and water also play a role. Due to their lower heat of evaporation, organic solvents decrease the flame temperature to a far smaller degree than does water. The solvent – dependence of the temperature of a hydrogen – oxygen flame into which either aqueous or methyl alcohol solutions are sprayed, is shown in Fig. 6.4 [225,] [620]. It is likewise obvious that the flame temperature depends on the amount of liquid introduced into it in a unit of time. Baker and Vallee [39] derived empirical equations for calculating the relation between temperature and the amount of water introduced into a flame in the form of an aerosol.

6.4 The flame spectrum

The emission of the flame may in some cases intefere in flame-photometric analysis. Of course, the so-called colourless flame of Bunsen or MECKER burners does not emit any significant spectrum with the majority of combustible mixtures, with the exception of intense OH bands. The spectrum of the outer and inner cones of an acetylene–air flame is illustrated in Fig. 6.5. The inner cone is used only rarely for flame-photometric purposes, as it emits an intense spectrum of OH, CH, CN, and C_2 molecules, depending on the gas mixture employed [299], [564].

The way in which CH or C_2 molecules are formed is nuclear, since all reactions by means of which these molecules might be formed are strictly endothermic. The presence of these molecules in the flame mainly indicates the vastly complicated character of the processes which take place in flames. The hypothesis has been published [298] that particles polymerise temporarily, and the molecules mentioned are formed by the disintegration of these polymers. As mentioned already, the intensity of this molecular radiation does not correspond to the true flame temperature and it cannot be used to measure this quantity.

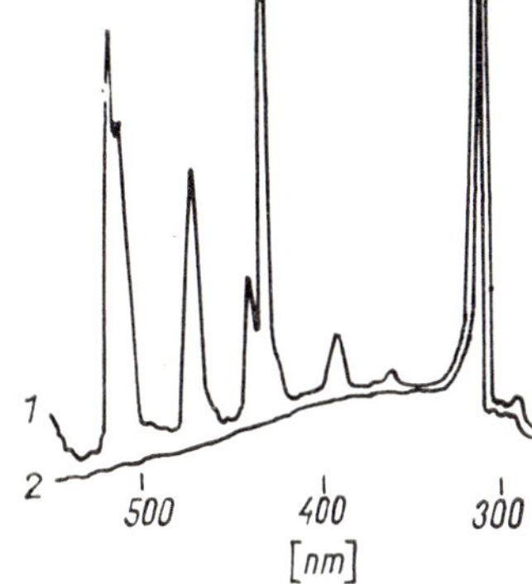

Fig. 6.5. The spectrum of an acetylene-air flame. The record was obtained with a Model CF-4 spectrophotometer by Optica, Milano

1 — spectrum of the inner cone,
2 — spectrum of the outer cone

The radiated spectrum also indicates that free radicals exist in flames. Frequently these radicals form very stable compounds with metals introduced into the flame, for example OH compounds with alkaline earths, lithium, and copper [712]. These molecular spectra often are intense enough to be used for analytical purposes instead of atomic spectral lines.

6.5 The mechanism of excitation in flames

The mechanism of excitation in flames has not yet been explained in a sufficiently satisfactory manner [298]. The reason is that collisions with electrons cannot be accepted as excitation processes, since the number of electrons in the flame is rather low (even in the reaction zone, where they are present in a substantially larger concentration compared with other parts of the flame), so that they cannot be a source of relatively intense radiation. The same certainly also applies to collisions of atoms with other atoms or molecules, because a large atom or molecule cannot transmit its kinetic energy to an electron. The basic possibilities therefore are conversion of vibrational or rotational energy on collision into electronic excitation energy [16]. Robinson [644] mentions a relationship between the radiation intensity of metal atoms introduced into the flame and the intensity of ultraviolet radiation emitted by the flame proper, basing this on observations in oxygen–hydrogen and cyanogen flames. Robinson derives the intensity of this radiation from spectral bands of OH particles in the oxygen–hydrogen flame (around 300 nm) and from the relatively intense radiation of shorter wavelengths in the cyanogen–oxygen flame. He concludes that the ultraviolet radiation formed in the flame is evidently converted to radiation of longer wavelength. Another possibility which deserves to be considered is chemiluminescence, specially in excitation of lines with a great excitation energy. Chemiluminescence, i.e. conversion of chemical into electronic excitation energy, is ascribed for example in carbon–containing flames to the reaction [318]

$$\mathrm{C} + \mathrm{MO} \quad = \quad \mathrm{CO} + \mathrm{M}^*$$

the overall energy balance of which is, in eV

$$\Delta H^\circ = -11.1 + D_{\mathrm{MO}} + E$$

where ΔH° is the standard enthalpy change of the reaction,

C – carbon atom,

M^* – excited metal atom,

D_{MO} – dissociation energy of the oxide, and

E – excitation energy of the respective line.

In flames poor in carbon and in flames in which hydrogen is used as fuel, the following reaction plays a major role

$$H_3O^+ + MO + e^- \quad = \quad M^* + H_2O + OH$$

with the overall balance of

$$\Delta H^\circ = -13.1 + D_{MO} + E$$

Thus the radiation of elements having a great excitation energy is bound to the reaction zone of the flame [848] and is rather difficult to study. Nonetheless, chemiluminescence has been proved to exist in the case of a number of elements [104], [180], [248], [317].

7. Inflow of the sample into the flame

7.1 Possibilities of introducing the sample into the flame

As mentioned already, the flame temperature is frequently insufficient for perfect evaporation of a solid sample. For this reason, the sample solution is usually dispersed and introduced into the flame in the form of an aerosol. Thus the sample, finely dispersed in particles several microns in size, evaporates in the flame in a matter of several milliseconds, and dissociation and excitation of free atoms takes place. These then emit their characteristic spectra.

Experiments have also been carried out with the introduction of a finely ground solid sample into the flame. The advantage of this procedure is the fact that the sample need not be dissolved, but the demands on mechanical sample preparation are greater and the precision of the determination is less than with the solution procedure.

Therefore, dispersion of a sample solution is used in the majority of cases in flame photometry. Inflow of the material analysed in the form of an aerosol is a guarantee of homogeneity, permitting relatively low temperatures to be used. The aerosol may be introduced into the flame either directly (using burners with so-called direct injection) or indirectly (by means of a mist chamber).

From the flame-photometric aspect we are mainly interested in the final result, i.e. intensity of emission or absorption of the respective line of the element analysed. Obviously, as long as the conditions in the flame remain constant, this intensity will be the greater the larger the amount of sample introduced into the flame in a unit of time, and the more finely the sample is dispersed. From the physical point of view, processes related to the inflow of the sample to the flame and excitation in the flame are by no means simple, being interrelated in different ways so that the relationship between intensity and the other parameters is not always quite obvious. The result is influenced already in the phase of preparation of the aerosol and its injection into the flame, as well as by the properties of the flame. The flame properties may depend on the amount of aerosol injected and on its composition, especially in the case of the direct-injection burner. Let us attempt to analyse the individual processes in order to elucidate the various relationships.

7.2 **Dispersion**

Nowadays, pneumatic dispersion devices are used conventionally in flame photometry. The main parts of the devices are the suction capillary, through which the solution to be dispersed passes, and a pressure nozzle delivering a stream of gas under pressure. Generally this gas is the oxidant. The suction capillary and the nozzle are usually arranged concentrically or perpendicular to each other. Sometimes the efficiency of the dispersion device is enhanced by placing an impact body in the path of the dispersed droplets, on which the droplets disintegrate, increasing the fineness of the aerosol obtained.

Other dispersion devices which deserve to be mentioned are the electrostatic type [737], in which the solution is dispersed by the effect of an electrostatic field of 5 to 15 kV, and the ultrasonic type [209]. However, these devices are not generally being used practically, and therefore we shall limit our following discussion to the operation of pneumatic dispersing devices. Their design is described in section 8.3.

The theory of dispersion is, in principle, based on the assumption that a liquid when forced out of the suction capillary forms ribbons under the influence of a gas stream, to which the RAYLEIGH theory of disintegration of a liquid column [127] may be applied. In the first phase, the liquid flows through the capillary and reaches its mouth. In the second phase, a rapid gas stream tears ribbons off the liquid stream. The stability of these ribbons is determined by the RAYLEIGH criterion, i.e. their length cannot be greater than their circumference. If this length is exceeded, the ribbons are torn away to form freely flying droplets which are capable of separating into smaller droplets in the course of their flight.

7.3 The size of aerosol particles

The size of the aerosol droplets formed is not uniform, exhibiting a roughly normal logarithmic distribution. In the majority of cases, the range of radii is approximately two orders of magnitude. An aerosol with this distribution of particle radii is generally characterised by some mean particle radius value, selected in such a way that if the real aerosol with its specific particle size distribution were replaced by a monodisperse aerosol with particles of a single size, the investigated properties of the aerosol should not change. However, it is no simple matter to find this value, since four quantities must be compared: overall volume, overall surface, overall number of particles, and overall sum of radii. Only two quantities at a time can be equal with both aerosols. From the possible six combinations, equality of the overall volume and overall surface area is usually selected according to SAUTER [666]. This choice is influenced by the fact that the majority of studies of liquid aerosols are oriented to problems of liquid fuel combustion. For flame photometry, however, it would be more purposeful to determined the arithmetic mean droplet radius, since coagulation of the particles as well as the rate of their evaporation both depend on their radius (see the following). SAUTER's radius is somewhat larger than the corresponding mean droplet radius value, since the volume of large drops has a greater effect on this value than their radius. The relationships between radii defined in different ways are given in the literature [725].

Quantitative calculation of the droplet size of the aerosol being formed from the dispersion parameters is rather difficult. The result depends, on the one hand, on the shape and arrangement of the dispersion device, and on the other hand on the physical properties of the liquid and gas in question. With a given apparatus, correctly adjusted, the properties of the dispersion device and the physical properties of the gas will be constant. This relationship may, therefore, be expressed by means of the equation

$$F(r, v, \varrho, \eta, \sigma) = 0 \tag{7.1}$$

where r is the droplet radius,

v – the gas flow velocity,
ϱ – the specific gravity of the liquid,
η – its viscosity, and
σ – its surface tension.

One of the important dimensionless arguments of this function is the WEBER number which determines the size of a flying droplet

$$We = v^2 \varrho r / \sigma \tag{7.2}$$

where v again is the relative velocity of droplet and gas. With slow impact of the gas stream onto the droplet, the Weber number may achieve a maximum value of 10; with rapid impact a maximum of 6. When this value is exceeded the droplet disintegrates [8]. Therefore, it is obvious from the WEBER number that aerosol particles will be the smaller the greater the relative gas velocity. With rising gas velocity the droplet radius decreases rapidly at first, later approaching asymptotically a certain limiting value. According to SCHEUBEL this minimum achievable radius is 6 microns in the case of water at velocities of 100 to 120 m/s. NUKIJAMA and TANASEWA found a value of 5 microns at a velocity of 330 m/s.

Although viscosity is not involved in the WEBER number, it is still significant to some extent. Viscosity of the solution slows down the deformation of droplets which necessarily precedes their disintegration. Thus it may happen that the droplet will leave the critical space in which the WEBER number is exceeded before it has been deformed enough, and disintegration will not take place. NUKIJAMA and TANASEWA [548] derived an empirical equation for the size of aerosol droplets, including the influence of solution viscosity:

$$2r = \frac{58{,}500}{v}\left(\frac{\sigma}{\varrho}\right)^{0.5} + 597\left(\frac{\eta}{\sqrt{\sigma\varrho}}\right)^{0.45}\left(1000\frac{Q_1}{Q_2}\right)^{1.5} \tag{7.3}$$

where Q_1/Q_2 is the ratio of the liquid volume Q_1 dispersed in a gas volume Q_2. The above relation applies to the range of $\varrho = 0.7$ to $1.2\ \mathrm{g\,.\,cm^{-3}}$, $\sigma = 19$ to $73\ \mathrm{dyne\,.\,cm^{-1}}$, $\eta = 0.003$ to 0.5 poise and $v <$ velocity of sound ($\mathrm{cm\ s^{-1}}$). If the ratio Q_1/Q_2 is less than 1/3000, the second term on the right-hand side may be neglected, in whi chcase the influence of solution viscosity on the aerosol droplet size is eliminated. With commercial instruments, however, the ratio of disperse liquid to gas passed through is usually larger. For example, with the Zeiss Model III instrument this value is roughly 1/2000. From equation (7.3) it also follows that the size of aerosol droplets decreases with the decreasing surface tension. Therefore, the droplet size distribution shifts in the direction of smaller radii when organic solvents are used, since these all have surface tension values lower than that of water [180]. A number of authors have attempted to increase the fineness of aqueous aerosols, and thus also the emission intensity, by adding surfactants. Some authors observed no influence [278], although they decreased the surface tension of the aqueous solution from 72 to 32.3 dyne cm^{-1} by adding 0.1 % of a surfactant. On the other hand, other authors [74] found a distinct emission peak with a surfactant concentration of 5 %. In our opinion, the surface tension of the solution in the equilibrium state, i.e. when the reagent is concentrated on the liquid surface, cannot be applied in this case. Rather, the surface tension of a homogeneous mixture should be considered, since the rate of formation of the droplets is comparable to the rate of migration of the surfactant. Therefore these reagents only become effective with larger concentrations. With even greater concentrations, on the other hand, surfactants may have an unfavourable influence, since they block the surface of the droplets, thus slowing down the process of evaporation and supporting the coagulation of the droplets [603].

Although viscosity has no substantial influence on the size of the aerosol droplets formed, yet it has a great degree of influence on the amount of solution dispersed. The Poiseuille law applies to laminar flow of a liquid through a capillary:

$$Q_1 = \frac{\pi R^4 P t}{8 \eta l}, \tag{7.4}$$

where Q_1 is the liquid volume passing through the capillary within a time t, R is the capillary radius, P the pressure difference at the ends ofthe capillary, l the capillary length, and η the dynamic internal friction coefficient. Since a number of authors observed an inverse relationship between viscosity and the amount of liquid flowing through the capillary and dispersed, it may be stated that the flow-rate in the capillary corresponds to the POISEUILLE law. WINEFORDNER and LATZ [820] found, by means of accurate pressure and capillary radius measurement, that deviations from the POISEUILLE law occur with low-viscosity liquids.

Since the majority of mixtures of organic solvents (alcohols and ketones) with water exhibit a viscosity maximum at a certain concentration, a minimum of the amount of solution dispersed is found in this region. Fig. 7.1 shows the relationship between the amount of solution dispersed and the composition of the solution for a number of organic solvents, as measured by NEEB [542]. The amount of solution dispersed also de-

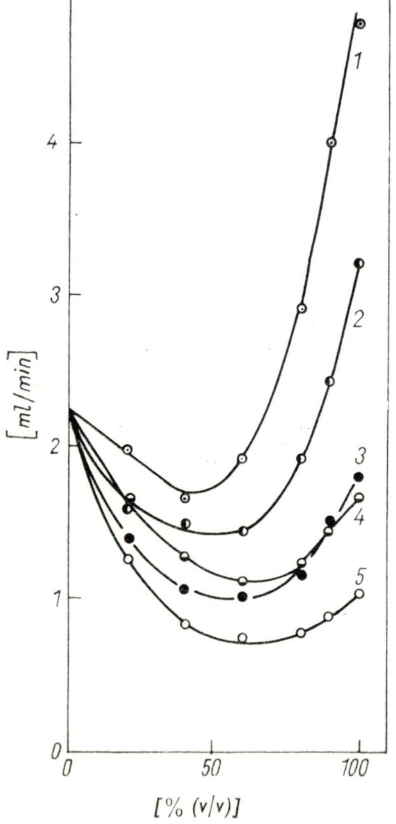

Fig. 7.1. Dependence of the rate of dispersion on the solvent composition for mixtures of water and
1 — acetone, 2 — methanol, 3 — ethanol, 4 — dioxane, 5 — isopropanol

Table 7.1. AEROSOL SEDIMENTATION

Particle diameter microns	Dropping rate mm/s
200	1200
10	3
5	0.7
1	0.003

pends, in the case of concentric dispersion devices, on surface tension [619], although to a lesser extent.

Starting at the moment when the aerosol first forms, two opposite processes begin to take place. One is evaporation of the solvent and decrease of the size of the droplets, the other process is sedimentation and coagulation of the particles. Specific experimental conditions decide which of the processes will prevail.

When a direct-injection burner is used the aerosol passes through a zone in which gases are heated, a reaction layer, and the flame, so that coagulation of the particles practically does not take place. When a mist chamber is used the aerosol has rather a long way to go before entering the flame, so that particle sedimentation and coagulation may take place to some extent. Let us now investigate these processes in more detail.

7.4 Sedimentation of aerosol particles

It is mainly the large droplets that are separated by sedimentation, so that the aerosol is rather fine when entering the flame. The Stokes law applies to the motion of particles in a gravitational field, and the maximum size of particles entering the flame can be derived by substituting the respective physical properties of the solution. Table 7.1 shows values found for an aqueous aerosol in air, according to CUNNINGHAM [154]. It is seen from these values that the diameter of droplets which enter the flame is not much more than 10 microns.

7.5 Aerosol coagulation

Further loss of the dispersed solution is due to aerosol coagulation. Coagulation takes place due to the impact of droplets on the walls of the mist chamber and also by mutual collisions and amalgamation of the droplets, which thus may exceed the size limit permitted by sedimentation.

The overall rate of coagulation therefore depends on the number and size of the particles, as well as on the size and shape of the mist chamber and the gas flow in it. Coagulation due to collisions of the droplets with the wall may be lessened by means of suitable adjustments, e.g. causing the gas in the mist chamber to execute a turbulent rotating motion [368]. Loss by coagulation may also be decreased by increasing the size of the mist chamber, although this is done at the expense of prolonging the time needed to stabilise the radiation of the element studied in the flame. Therefore the optimum size of the mist chamber also depends on the other parameters of the apparatus [370].

It is not clear whether coagulation is also influenced by an electrostatic charge which the particles may acquire on being dispersed. In this case, the charge would decrease coagulation by mutual electrical repulsion of the particles.

7.6 Evaporation in the mist chamber

Immediately after the aerosol has been formed, evaporation of the solvent sets in and the droplets decrease in size. Thus, the effect of this process is the reverse of coagulation. The rate of evaporation of the droplets is limited by the rate of diffusion of the molecules evaporated into the surrounding gas, and does not therefore depend on the surface area of the droplet, but only on its radius. LANGMUIR derived the following equation for the stationary state

$$\frac{\mathrm{d}m}{\mathrm{d}t} = \frac{4\pi DMpr}{\boldsymbol{R}T}, \tag{7.5}$$

where $\mathrm{d}m$ is the loss of the liquid in the time $\mathrm{d}t$, D is the diffusion coefficient of the solvent in the gas phase, M the molecular weight of the solvent, $\boldsymbol{R}$ the gas constant, T the absolute temperature, p the solvent vapour pressure at the given temperature and r is the droplet radius. With increase in motion of the droplet relative to the environment the rate of evaporation should increase. With the conditions prevailing in the mist chamber, where droplets are being carried away with the gas stream, the simple LANGMUIR equation may be used. Likewise the contribution of the increased tension over curved surfaces is negligible [219].

In flame photometry salt solutions are used, which according to the RAOULT law decrease the solvent vapour pressure. In this case, the partial pressure of the saturated solution must be substituted in the LANGMUIR equation, although the solution is more dilute in the initial phases [626]. The reason is that the dissolved salt concentrates rapidly at the surface and crystallises. The solution penetrates through this layer and the solvent evaporates. Solid aerosol particles entering the flame therefore take the shape of empty balls [503].

The case of droplet evaporation is somewhat more complicated, since the partial pressure of the solvent in the gas phase also participates in the process. Thus the difference of vapour pressure and partial pressure in the gas phase must be substituted into the LANGMUIR equation

$$\frac{dm}{dt} = \frac{4\pi DMr}{RT}(p_2 - p_1), \tag{7.6}$$

where p_1 is the partial pressure of the solvent in the mist chamber and p_2 the solvent vapour pressure at the given temperature.

Assuming that the aerosol is dispersed uniformly throughout the volume Q_2, expressing this partial pressure by the equation of state and relating the equation thus modified to the entire aerosol, the overall rate of evaporation of the aerosol is obtained in terms of solvent volumes

$$dV/dt = KV^{1/3}(V_{sat} - V_0 + V) = KV^{1/3}(V_{sat} - V_{evap}), \tag{7.7}$$

where V is the overall volume of the aerosol droplets at the time t, V_0 the volume of aerosol droplets on dispersion at the time $t = 0$, V_{sat} the solvent volume which can evaporate in a given gas volume Q_2, V_{evap} the volume of evaporated solvent in the time t, and K a constant given by the relation

$$K = \frac{4\pi D}{Q_2}\left(\frac{3}{4\pi}\right)^{1/3} n^{2/3}. \tag{7.8}$$

This constant K includes the diffusion coefficient of solvent molecules, D, and the number of drops, n, which in turn includes the influence of the drop radius on the rate of evaporation. In general, the constant K does not vary much with different solvents.

The course of evaporation of the aerosol mainly depends on the expression $V_{sat} - V_0$, which takes positive as well as negative values. Two principal cases must be distinguished. The one where the amount of solvent dispersed is greater than the amount capable of evaporating in the given volume ($V_{sat} < V_0$) is more common. In this case the rate of evaporation quickly decreases to zero and coagulation plays a major role. Therefore, there is no purpose in excessively increasing the amount of liquid dispersed. The other case sets in when highly volatile solvents are used, e.g. ketones or ethers. Here, $V_{sat} > V_0$ and the rate of evaporation of the aerosol is far greater. The increased amount of dispersed liquid also causes a relatively greater overall amount of evaporated free atoms to be present in the flame.

Thus, when liquids of higher vapour pressure are used, rapid evaporation takes place, coagulation decreases, and the amount of the substance introduced into the flame is greater. This advantage of a higher yield, however, is accompanied by other phenomena in the flame: the volume of the flame generally increases and its temperature rises.

On dispersion of the solvent, the temperature in the mist chamber decreases, in part to expansion of the gas on leaving the nozzle, as a consequence of the JOULE—THOMSON effect (a decrease of 1 to 2 °C) [11], but also due to the effect of evaporation of the solvent. Table 7.2 presents temperature values measured in the mist chamber of the Zeiss Model III instrument with different solvents dispersed. The initial temperature in these measurements was 22.4 °C.

Table 7.2. TEMPERATURE IN THE MIST CHAMBER

Solvent	Temperature °C	Solvent	Temperature, °C
Water	16.4	n-Propanol	14.2
Methanol	2.4	n-Butanol	18.2
Ethanol	8.6	Amyl alcohol	19.0

With the decreasing temperature, the vapour pressure of the solvent and rate of evaporation also decrease. Therefore it is sometimes recommended to heat the mist chamber, or to preheat the pressurised gas [734]. This latter procedure is more advantageous, since with the former method the danger of crystallisation of salts on the dispersion nozzles increases, incurring the danger of defects.

The above two processes, i.e. coagulation and sedimentation on the one hand, and evaporation on the other, influence the overall efficiency of the dispersion device. This term is understood to mean the proportion of the dispersed solution which reaches the flame. Obviously, efficiency may be improved by decreasing the degree of coagulation and sedimentation, which may be achieved for example by decreasing the size of aerosol droplets. A similar effect on efficiency may be obtained by speeding up the evaporation process by using solvents of higher vapour pressure.

With elements which are completely dissociated in the flame (e.g. Cu, Ag, Zn, Cd) the influence of organic solvents, the vapour pressure of which is always greater than that of water, is mainly due to the increased amount of the aerosol entering the flame.

7.7 Evaporation in the flame

In direct-injection burners the aerosol enters the flame in the form of a liquid: when a mist-chamber is used, a part of the aerosol converts to a solid before it enters the flame. In the flame, however, the solvent evaporates immediately, most of it even before the reaction zone, so that the two cases become very similar in character. It must be remembered that with direct-injection burners the solvent has a great influence on the conditions prevailing in the flame. Evaporation of solid particles is far slower, and its duration is comparable with the time of flight of the particles through the flame. The rate of evaporation of solid particles, which are substantially smaller than aerosol droplets, and smaller than the mean free path of the evaporated molecules, is given by an equation which applies to evaporation in vacuo [825]

$$\frac{\mathrm{d}m}{\mathrm{d}t} = 4\pi r^2 p \sqrt{M/2\pi \boldsymbol{R}T} \tag{7.9}$$

The notation used is the same as in equation (7.5).

Thus the rate of evaporation does not depend on the temperature, vapour pressure or boiling point of the compound, or the surface area of the particles. As long as the boiling point is lower than the flame temperature, the vapour pressure will be greater than 1 atmosphere and evaporation will take place immediately. This is the case with conventional alkali metal salts. When, however, the boiling point is higher than the flame temperature, the vapour pressure will be less, and only partial evaporation can take place within the time of flight of the particle through the flame. In this case the size, or rather surface area of the particles will play a major role.

Obviously, evaporation usually is preceded by fusion of the salts, which may cause various chemical interactions between the components of the melt or with the combustion gases. An important factor is the formation of compounds of alkaline earths with some anions (PO_4^{3-}, SO_4^{2-}), or with polyvalent metal oxides (Fe, Al, Ti, Cr), which volatilise with difficulty.

Fractional distillation may also take place, with the disadvantageous effect that when the more volatile component has evaporated the solid particles decompose to form finer ones, or they acquire a sponge-like structure. In both cases the surface area of the particle grows and evaporation of the less volatile component is speeded up.

7.8 The significance of theoretical observations for practical analysis

In the preceding sections, we have briefly discussed the qualitative aspects of the basic processes which underlie the method of flame photometry. It is true that the relation between the intensity of emission of a certain element and the number of atoms introduced into the flame can be calculated exactly from the flame temperature, its geometric dimensions, and the composition of the combustion gases, but this is only possible in special cases and with the use of a computer. In practical analysis too many parameters vary, since a number of physicochemical processes are involved. These generally have complicated kinetics, comparable in duration to the flight of atoms through the flame, so that such investigations are mady very difficult. The expression physico-chemical processes is here understood to include the rate of evaporation of the solvent, formation of solids, rate of evaporation of their molecules, dissociation to free atoms and radicals, and chemical reactions with other radicals or ions occurring in the flame. Stationary states differ therefore in different parts of the flame, and much also depends on the rate of flow of gases in the flame. The intensity of the radiation emitted and, to some degree, of its absorption is therefore the result of a complex of relationships. For this reason, different authors using different instruments will frequently come to different, often controversial, conclusions.

Thus, it is probable that flame photometry will continue to be a comparison method, requiring the preparation of reference standards.

PART III

PRACTICAL

8. Instruments and equipment

At present, some 35 different types of flame photometers are being manufactured commercially [302]. These differ to varying extents from each other in design. Moreover, the design of at least an equal number of instruments has been suggested and published in the scientific literature, not to speak of improvements and modifications of the individual parts of the commercial instruments. In principle, however, flame photometers include basic components which occur in different modifications in nearly all types.

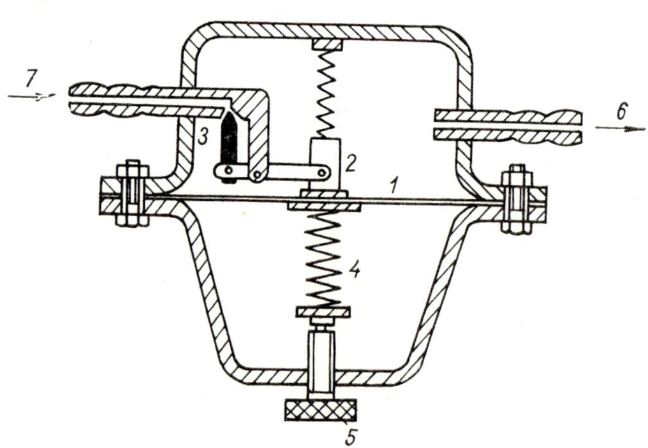

Fig. 8.1. Diagram of a reduction valve

1 — membrane, 2 — lever which controls the needle valve, 3 — needle valve, 4 — adjustment spring, 5 — control screw, 6 — outlet of gas at reduced pressure, 7 — inlet of highpressure gas

A sketch of a flame photometer for emission measurements was shown in Fig. 2.1. It consists of the following parts:

1. a pressure bottle containing gases and a pressure regulator (Fig. 8.1), dispersion devices and mist chambers
2. the burner
3. optical equipment for isolating the radiation studied
4. detection equipment
5. a measuring instrument

In the case of instruments for absorption measurement, a source of monochromatic radiation must also be included together with the respective feed source. Besides these main parts, some instruments are fitted with other facilities.

The characteristic properties of the individual parts of the apparatus will be outlined in the following.

8.1 Gases, equipment for their storage and purification, flow into the apparatus and pressure regulation

As mentioned already, various substances are used in flame photometry as fuels and oxidants. These are mostly gases. Fuels which are liquid under normal laboratory conditions (e.g. benzene or acetone [77], [585]) have been suggested for use in isolated cases only.

Individual gases differ from each other in the manner of preparation as well as of storage and regulation. For practical purposes, the following demands must be satisfied:

1. the gases must be easily available,
2. they must not be too expensive,
3. they must permit pressure regulation,
4. their composition must not vary in the course of the measurement.

When working with some gases, certain safety precautions must be maintained: these will be mentioned in the respective section of this chapter or in the chapter devoted to technical work with the flame photometer.

8.1.1 Air

Air is one of the oxidants most often used. It is supplied in pressure bottles or it may be supplied by suitable compression equipment. Pressure bottles containing air are usually marked with aluminium paint on the upper part. A gas supply fed from pressure bottles guarantees a high degree of gas pressure stability, and it is easy to control. It requires, however, that bottles be transported frequently, since the amount of gas they contain is insufficient for a prolonged period of operation (shorter, for example, than in the case of acetylene-containing bottles). Air is usually supplied in bottles of 40 l volume, which contain slightly more than 6 m^3 gas at a pressure of 150 at. The air supplied from these bottles is sufficiently pure. Only with old, corroded bottles is there some danger of contamination with rust, which the pressurised gas may carry away in fine particles. Fine dust may also be contained in gas, which therefore must be purified before use, for example by passing through a filter of glass wool or a glass sinter plate. Obviously, a filter must be selected which will not liberate any solid particles of the filter material, which might further contaminate the gas.

Bottles sometimes are filled with air in the same apparatus which is also used to fill oxygen bottles: in this case, the air may have an increased oxygen content. Such a change in the composition of the combustion gases may be the cause of an explosion, or at least of altered combustion conditions. Therefore, it is essential before connecting a new bottle onto the apparatus to make sure that the air is not enriched in oxygen (e.g. by means of a smoldering match).

The supply of air from a compressor is more convenient. With some types of instruments this equipment is built directly into the instrument. Pumps of the so-called dry type are very suitable. A disadvantage of this procedure is the fact that the air pressure may vary rather greatly, making a correct analysis impossible. Therefore, a device must be included in the air inlet to compensate for pressure variations. At the same time, the pressure must be checked with a manometer.

When a compressor is used, the air may be contaminated with oil droplets, and it is useful therefore to include in the apparatus a device for removing these droplets. Most suitable is a glass sinter or a cotton-wool filter.

The gas inlet, from a bottle as well as from a compressor, must be regulated in some way. Flame photometers conventionally operate with overpressures of up to 1 at., while pressure bottles are filled to

150 at. When the gas is taken from a bottle, a reduction valve is used. For every type of gas there is a different type of valve, which may not be interchanged. Individual valve types differ from each other in the size and direction of the thread which joins them to the bottle, and they are moreover painted in a colour which is reserved for a given gas. The principle of a reduction valve is shown diagrammatically in Fig. 8.1. A suitable pressure is set by means of a membrane connected to the valve proper. If the operating pressure is low (up to 1 kg/cm^2, or 14 psi) the small high-pressure membrane does not allow reliable regulation. Therefore, two valves are usually connected in series. The first is a high-pressure valve, regulating the pressure down to values of the order of 10 kg/cm^2 (142 psi), the second has a larger membrane and can be set finely to regulate the pressure down to operational values. With lower pressures (from a laboratory compressor) the gas is controlled by increasing the resistance in the inflow to the burner, for example by narrowing down the inlet profile and fine control is achieved by removing the excess air through an outlet, e.g. a rubber tube fitted with a screw clamp.

8.1.2 Oxygen

If a higher flame temperature is needed, oxygen is usually selected to replace air, which carries nitrogen as ballast. Oxygen is taken exclusively from pressure bottles, usually marked on the upper part in blue. Oxygen in bottles is usually very pure, its content not decreasing below 99 %. It is obtained by fractional distillation of liquefied air. Like air, oxygen may be contaminated with rust if it is put into old corroded bottles. Therefore, new bottles should be used for flame photometric purposes, or some bottles should be reserved exclusively for this purpose. Oxygen is purified in the same way as air. To measure its pressure, steel manometers are fitted in the apparatus. Oxygen, like air, is fed into the apparatus through rubber tubes some 0.7 cm in diameter. With higher pressures it is sometimes preferable to use so-called armoured tubes with cord inlays instead of conventional tubes.

8.1.3 Acetylene

Acetylene is the most conventional fuel in flame photometry. It was used practically for the first time by LUNDEGARDH, and is used in mixtures with air as well as with oxygen. Its preparation in the laboratory by means of decomposition of calcium carbide is very impractical. Acetylene usually is supplied in pressure bottles marked in yellow. Differing from oxygen or air, acetylene is not simply compressed in the bottle: it is dissolved in acetone. The reason is that when compressed, acetylene has a great tendency to self-induced decomposition and explosion. Pressure bottles used for acetylene are filled with a porous material – diatomaceous earth and charcoal or some similar material – impregnated with acetone. The solubility of acetylene in acetone is great: under normal conditions, with a filling pressure of 150 at, one litre dissolves roughly 375 l acetylene. The packing removes the danger of acetone flowing out when the bottle is inclined at an unsuitable angle, at the same time hindering explosion in case of an unexpected back-flash of the flame.

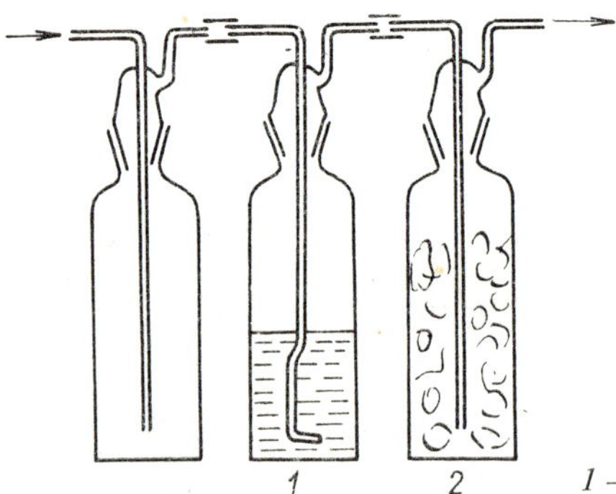

Fig. 8.2. Diagram of equipment for purification of acetylene
1 — conc. sulphuric acid, 2 — glass wool

When acetylene is being taken from the bottle, however, it should never be in a horizontal position. The packing takes up roughly 25 % of the bottle's volume, acetone another 38 %. Its volume increases some 29 % when the acetylene has been dissolved in it. For reasons of safety the remaining volume is left free. The conventional commercial bottle of 40 l volume contains approximately 5.5 m^3 dissolved acetylene. The pressure within the bottle depends substantially on the laboratory temperature, it may rise to 25 kg/cm^2. Acetylene should not be removed completely from the bottle, because at low pressures it contains substantial quantities of acetone: this lowers the flame temperature. Acetylene is frequently contaminated, mainly by hydrogen phosphide which imparts a very disadvantageous colour to the flame. For this reason, acetylene must be purified before use. It is allowed to pass through a special filter of RASCHIG rings impregnated with ferric chloride, ferric oxide, silver oxide and potassium permanganate, or it is passed through concentrated sulphuric acid [510]. This equipment, composed of three washing bottles, is shown in Fig. 8.2.

The pressure of acetylene entering the instrument is measured with a manometer filled with a liquid in which acetylene is insoluble. Most often this is weakly alkaline water or dibutyl phthalate.

Acetylene is a relatively poisonous, dangerous gas. However, it is rather easily recognised if it leaks from inlet tubes or from the pressure bottle, by the characteristic hydrogen phosphide smell which always accompanies acetylene gas. The explosion level of acetylene is even lower than that of town gas.

8.1.4 Hydrogen

Hydrogen is one of the most frequently employed fuel gases. Its advantage is the fact that the hydrogen flame has no spectrum of its own. Hydrogen is exclusively taken from pressure bottles which are usually marked in red: they are usually filled to 150 at. For flame-photometric purposes it need not be purified preliminarily. Its use is subject in principle to the same precautions as that of oxygen. With air it forms an explosive mixture and therefore great care must be exercised when hydrogen is employed. Since hydrogen has no smell, leaks are more difficult to find than in the case of acetylene. However, hydrogen is lighter than air and stays near the roof in a room, which fact somewhat decreases the explosion hazard. Armoured tubes (i.e. tubes with cord inlay) must be used for hydrogen tubing, and the entire apparatus must be tested thoroughly for leaks.

8.1.5 Propane and butane

These gases are supplied in pressure bottles marked on the top in orange. Special safety precautions apply when working with the two gases and their mixtures. Propane and butane are both heavier than air and remain near the laboratory floor. Safety rules demand, therefore, that there be openings in the lower parts of laboratory doors for the gases to escape from the room. The pressure bottle should be located outside the laboratory.

Their greatest disadvantage, mainly that of butane alone, is that they are never very pure. Propane and butane commonly contain other related compounds of somewhat different physical properties. Thus the properties of the gas vary while it is being taken from the bottle. At first the most readily volatile compounds leave the bottle; later it is mainly butane, and lastly the heavier compounds. The flame temperature may vary accordingly, and the experimental conditions are never defined accurately. It is more advantageous to use propane alone, which also is not usually obtained as a chemically pure gas, but is accompanied by compounds of closely related physical properties, so that these do not alter the flame temperature.

Both gases are introduced into the apparatus by means of rubber tubes, as in the case of acetylene. Preliminary purification is usually not required.

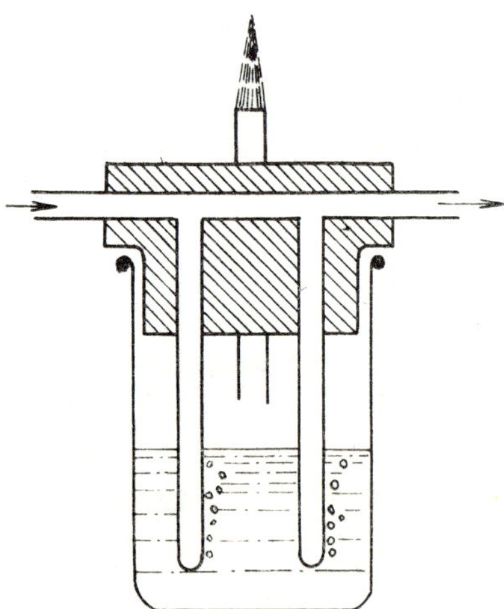

Fig. 8.3. Equipment for maintaining a constant pressure of town gas

8.1.6 Town gas

Town gas has been suggested for use in some determinations. Town gas is taken from the distribution piping, and exceptionally it may also be obtained in pressure bottles. Thus town gas is easily available in most laboratories. Its disadvantage is pressure oscillations, which however may be compensated to some degree by means of the equipment shown in Fig. 8.3. However, the operation of this equipment must be checked in the course of an analysis, because with large pressure variations the flame in which an excess of gas is burned may go out. Another disadvantage is frequent alterations of the gas composition. Products from different gasworks differ as much as the product of a single supplier varies during one work shift. An example of the composition of town gas is given in Table 8.1. Generally the hydrogen and carbon monoxide concentration varies substantially, obviously entailing changes in the flame temperature.

Table 8.1. COMPOSITION OF TYPICAL TOWN GAS

Component	%	Component	%
CO_2	4.0	H_2	51.3
C_mH_n	2.0	CH_4	17.0
O_2	0.2	N_2	4.0
CO	21.5		

8.1.7 Other heating mixtures

VALLEE and BAKER [773] published a study dealing with the use of cyanogen, which together with oxygen gives a high-temperature flame, close in temperature to the electric arc. This source of light allows spectra to be formed in which even the lines of elements difficult to excite appear. However, there are some disadvantages related to the use of this gas. It is a very poisonous gas, requiring special safety precautions. From the analytical point of view the determination of some elements is made difficult by the very intense spectrum of the flame in the violet and ultraviolet regions.

Recently, the nitrous oxide – acetylene flame is becoming more widely applied in atomic absorption techniques. This flame has rather a rich molecular spectrum, which however is no hindrance with instruments fitted with modulation equipment. This mixture has the advantage of a high flame temperature [815] while the possibility of using a burner with mist chamber is retained. An even higher flame temperature is obtained with the nitric oxide – acetylene flame [26], although some authors [707] believe that this latter flame will find no practical application, since nitric oxide is far more expensive than nitrous oxide and, moreover, it has unfavourable corrosive and toxic properties.

The use of perchloric acid fluoride mixed with hydrogen [693], [676] is also described in the literature. This mixture gives a flame nearly equal in temperature to that of cyanogen and oxygen. It has, however, found no practical application up to now. Roughly the same holds for the hydrogen-fluorine mixture [144], [819].

In an effort to achieve high temperatures, other possible sources of energy were also suggested, e.g. the combination of a flame and a spark discharge [390], [414], [468], [481], [647], a torch-like discharge [675], [747], [760] and a plasma burner [308].

8.2 Burners

A burner is a device which permits a gas mixture to burn with a stationary flame. In practical flame photometry, various types of burners are used, which may be classified roughly into two groups. The first includes those types which operate simply as a burner, while the second group includes direct-injection burners, which combine the function of a burner and of a dispersion device. The burners of the first group belong with instruments fitted with mist chambers, and give laminar or nearly laminar flames. Direct-injection burners give turbulent flames. The burner mouth must be designed in such a way as to permit the passage of gases in amounts needed for burning and for a stationary flame to form, and at the same time to prevent the flame from back-flashing.

8.2.1 Burners with mist chambers in their inlet

In these burners the fuel is introduced by a narrow nozzle and mixed with oxidant gas which carries an aerosol of the solution to be analysed. The path which the two gases are forced to travel must be large enough for the final flow to be laminar. Therefore, in some types of apparatus the fuel and oxidant gases are first led to a mixing chamber where they are mixed thoroughly before entering the burner proper.

The burner mouth is selected in such a way as to avoid back-flashing of the flame. With BUNSEN-type burners with open upper end the flow velocity must be roughly three times as large as the rate of combustion of the respective mixture. Therefore these burners are only used for mixtures with relatively slow rates of combustion.

A generally more secure operation is obtained with MECKER burners, which are fitted with a removable cap at the top. The cap has a number of holes, the diameter of which should not be greater than the so-called quenching diameter of the respective gas mixture. The principle of the function of this cap is the fact that the amount of heat conducted away by the cap walls is large enough to keep the gas from achieving the ignition temperature before the gas has left the channels in the cap. The values of quenching diameters for some of the more widely used mixtures are given in Table 8.2.

Table 8.2. QUENCHING DIAMETER VALUES

Gas mixture	Diameter, mm
Propane, butane — air	2.5
Acetylene — air	0.9
Hydrogen — oxygen	0.25

In the case of butterfly-type burners the burner mouth has the shape of a narrow slit, the width of which again must not be greater than the respective quenching diameter. These burners give long narrow flames and are mainly used in absorption flame photometry.

The advantage of the above systems is mainly their laminar flame, which, together with fine dispersion of the aerosol, secures good reproducibility of the analyses. The flow velocity of the gas is such that aerosol droplets evaporate on passage through the flame and mostly convert to the gaseous state while still in the reaction zone.

Some gases, mainly hydrogen and oxygen, which cannot very well be mixed in the burner, are conducted through separate tubes to meet at the burner outlet. A diagram of one such water-cooled burner is shown in Fig. 8.4.

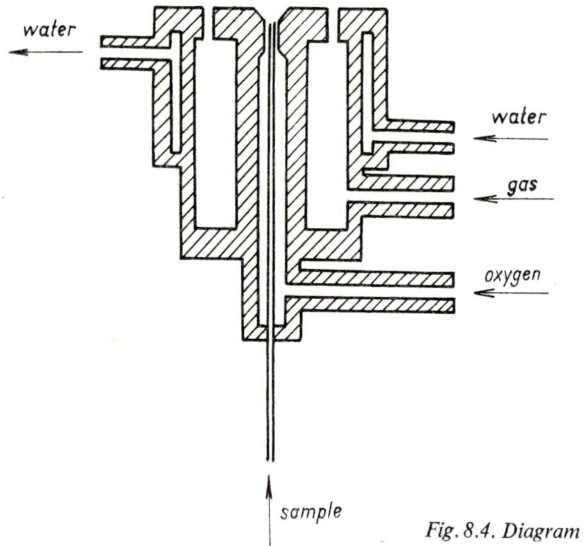

Fig. 8.4. Diagram of water-cooled burner

8.2.2 Direct-injection burners

A burner which at the same time acts as a dispersion device is shown in Fig. 8.5. The type most frequently employed is designed on BECKMAN'S principles and is therefore called the BECKMAN burner. It is used in instruments which have no mist chamber, and the solution to be analysed is led directly into the flame. The burner is formed by concentric tubes. The innermost one is the suction capillary, which is usually made of palladium, platinum, or an alloy of some platinum metals, or rarely of stainless steel. Capillaries of glass or quartz are unsuitable, because they are too fragile and might easily break in the centring operation, which must be done rather often with this type of burner. The oxidant – oxygen or air – flows through the annular space around the capillary, carrying away the solution from the capillary. The flow velocity of this gas must be chosen such as to permit an aerosol to be formed while not quenching the flame. The heating gas is introduced through the annular space at the circumference of the burner. With this burner also, dimensions of the inlet tubes and the suction capillary must be chosen suitably for every individual gas. Since oxygen is mostly used as oxidant, the rate of combustion is high. This again requires that the flow-rate of the two gases be adjusted accordingly. On correct operation of the burner the lower part of the flame does not rest on the burner, being usually located several millimetres above. The flame obtained is a turbulent one and it is accompanied by a strong roaring noise. Some types of these burners are fitted with cooling facilities. Their disadvantage is the fact that the flame temperature is influenced by the character of the solvent used, and that the size of aerosol particles is rather irregular. Large droplets will frequently fly through the flame without evaporating. However, there is the advantage of complete utilisation of the solution analysed and of a greater immediate concentration of the element analysed in the flame.

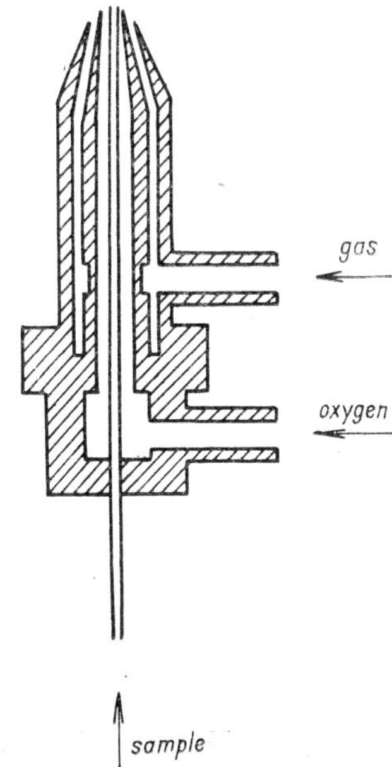

Fig. 8.5. BECKMAN *burner*

8.2.3 Other types of burners and equipment

Besides the conventional types already discussed, some special burners should also be mentioned. Firstly there is the burner for the cyanogen—oxygen flame [773], [788]. It is shown in Fig. 8.6. A burner was also

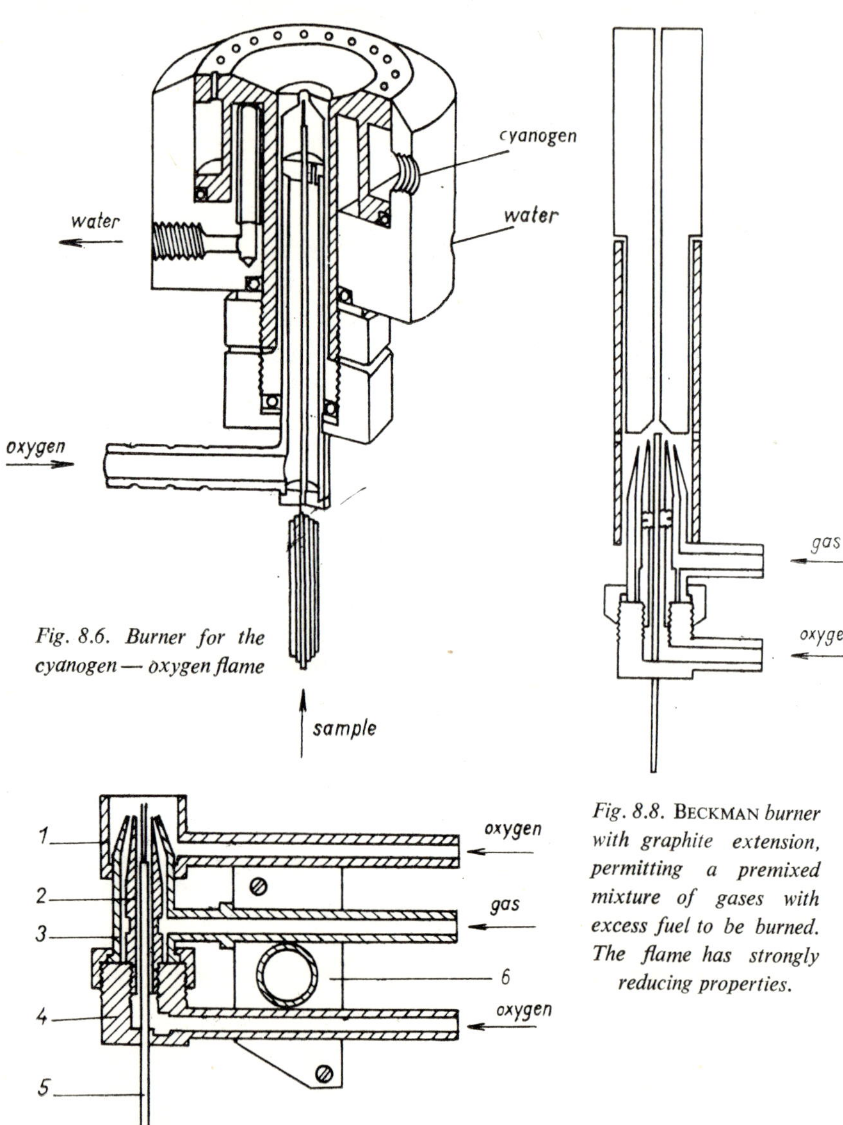

Fig. 8.6. Burner for the cyanogen — oxygen flame

Fig. 8.8. Beckman *burner with graphite extension, permitting a premixed mixture of gases with excess fuel to be burned. The flame has strongly reducing properties.*

Fig. 8.7. Modification of the Beckman *burner for combustion in a protective atmosphere*
1 — mantle for protective oxygen, 2 — for oxygen, 3 — for fuel, 4 — burner base, 5 — suction capillary, 6 — holder

proposed (Fig. 8.7) which permits combustion in a "protective atmosphere" of oxygen [48], [316], [488]. The equipment proposed by KNISELEY et al. [447] is a modification of the BECKMAN burner. Its principle is illustrated in Fig. 8.8.

In some instruments the burner is covered with a little chimney of cast iron, or at least a protective wire net. The purpose is improved safety of work. Many flames are quite colourless and the operator might easily burn himself by inadvertently coming close to the flame.

8.3 Dispersing devices

The dispersing device forms a fine aerosol of the solution containing the dissolved sample. The solution must be very finely dispersed for the size of aerosol droplets not to be greater than microns. This size of droplets is necessary as it permits a sufficient amount of the solution to be transported into the flame, facilitates evaporation of the solvent, as well as other processes taking place in the flame essential for a sufficient number of free atoms to be available in the flame for excitation or absorption.

It is clear, therefore, that the dispersing device is an important part of the instrument, and that the sensitivity and precision of the analysis depend on the design and regular operation of the device. Other demands also apply to the dispersing device. It must secure a high degree of reproducibility of the aerosol properties; it must be made of a non-corrosive material; its parameters must be selected so as to permit a suitable flowrate of the gas through the apparatus; and finally, its adjustment should not be too complicated.

Several types of dispersing devices were suggested and practically applied. They may be classified under the headings of self-propelled dispersion devices operating on the pneumatic principle, and those which depend on outside energy for the dispersion process.

8.3.1 Self-propelled dispersing devices

The mechanism of dispersion is described in detail in section 7.2. Here we shall only mention the basic principle. A gas flowing at high velocity sucks up a liquid and disperses it. With a device of this sort, therefore, the aerosol is prepared by pneumatic means, the propelling gas usually being air or oxygen. Some authors have also obtained very good results with dispersion devices in which acetylene was used as propellant [316], [477].

Self-propelled dispersion devices may be divided into two groups. In the first the suction capillary is placed perpendicular to the nozzle. An example of this group is the dispersion device of the Modell III Zeiss flame photometer, or in the flame-photometric adaptor of the Hilger UVISPEK spectrophotometer (Fig. 8.9). In the second group, the suction capillary and the pressure gas nozzle are concentric, again with two possible design principles: with the first the capillary juts out above the mouth of the pressure nozzle or is located in the same plane. The efficiency of the device depends on the relative positions of the two mouths, and the amount of gas passing through depends on the width of the annular space. This type of dispersion device is used, for example, in the BECKMAN direct-injection burner (Fig. 8.10a). In the second case (see Fig. 8.10b) the mouth of the suction capillary is below the level of the pressure nozzle. Dispersion devices of this kind are used, for example, in the Model SP 900 Unicam flame photometer. The amount of gas passing through depends, in this case, on the inner diameter of the pressure nozzle.

As already mentioned in the chapter on dispersion, it is advantageous for the overpressure of the dispersion gas to be ca. 0.9 at. The flow velocity of the pressure gas in the nozzle then achieves the velocity of sound and the efficiency of the dispersion device depends less on pressure variations.

The amount of liquid which passes through the device obviously depends on the inner diameter of the suction capillary (Table 8.3). Capillaries of larger diameter permit larger sample amounts to be dispersed within a unit of time, but the mean particle size of the aerosol formed is larger and the degree of utilization of the aerosol decreases. The more narrow the capillary, the finer the aerosol which is formed. The minimum capillary diameter is limited, however, since it is frequently fouled by mechanical impurities contained in the solutions. This causes variations in the type of aerosol produced. Therefore, inner capillary diameters are selected for practical purposes in the range of 0.2 to 0.6 mm.

The inner diameters of the suction capillary and dispersion nozzle also depend on the type of pressure gas employed. When this is air the amounts of gas delivered are usually very large, and the amount of liquid dispersed may also be increased without diminishing the proper-

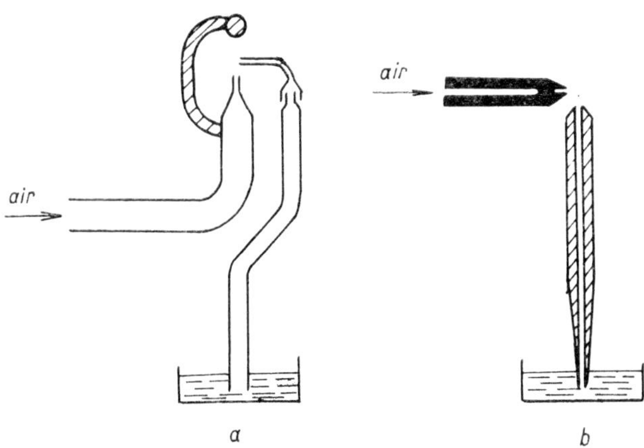

Fig. 8.9. Angle-type dispersion devices

a — Zeiss Model III flame photometer,
b — flame-photometric adaptor for the Hilger Uvispek

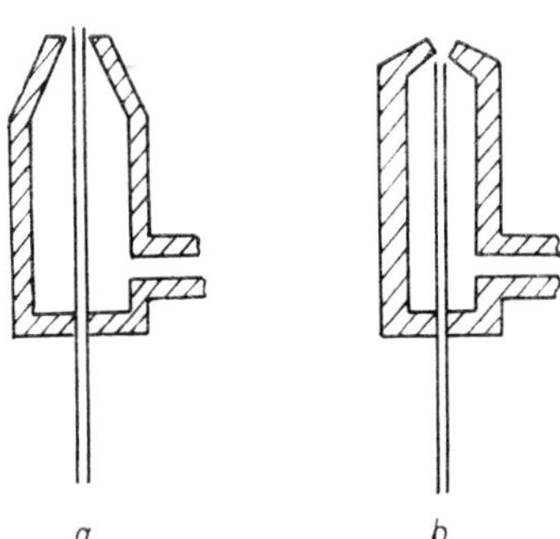

Fig. 8.10. Concentric dispersion devices

a — suction capillary above the nozzle level, b — suction capillary below the nozzle level

ties of the aerosol. On the other hand, oxygen, of which smaller amounts are needed for combustion, only permits a small amount of the solution to be dispersed in a unit of time. In this case, therefore, a suction capillary of smaller diameter should be used.

Table 8.3. RATE OF DISPERSION OF SOLUTIONS (ml/min) WITH DIFFERENT DIAMETERS OF THE SUCTION CAPILLARY

(Zeiss Model III flame photometer)

Solution composition	Suction capillary diameter	
	0.4 mm	0.6 mm
Water	8.1	12.2
20 % ethanol	6.85	9.75
40 % ethanol	5.95	8.5
60 % ethanol	5.8	8.4
80 % ethanol	6.1	8.7

Concentric dispersion devices are usually more efficient, but their adjustment is more difficult than with the other type. Special care must be taken that the suction capillary is exactly concentric with the pressure nozzle, and the distance of the capillary mouth from the pressure nozzle is also of great importance. The edges of the two mouths should be as sharp as possible and free of mechanical irregularities.

Angular-type dispersion devices are usually made of glass, which makes them corrosion-resistant. Capillaries for concentric dispersion devices are made of metal, e.g. stainless steel or alloys of platinum and iridium. Glass is unsuitable for this purpose, since the adjustment of the device necessitates high-strength materials.

The outer mantle of the concentric type of device may be of metal (e.g. in the case of direct-injection burners), of plastic, or of glass. Plastic mantles cannot be used for work with organic solvents, e.g. chloroform.

In the case of instruments fitted with mist chambers an impact body ("Prallkörper" in German) is sometimes placed in the path of the aerosol formed to improve its properties. It is difficult to decide whether this improvement takes place by disintegration of aerosol droplets in the impact, or whether the improvement is due to ultrasonic effects, which some authors believe to be related to the function of the impact body.

In absorption flame photometers, the same equipment is used as in the emission type of instrument. The diameter of the suction capillary and pressure nozzle is usually larger, however, because of the different design of the burner, which necessitates a larger amount of the fuel mixture.

8.3.2 Dispersion devices powered by an external source

These devices are of no great practical significance. Let us mention several proposed types only for the sake of completeness. One type was already suggested by BUNSEN, who introduced granulated zinc into an acid solution for analysis. Due to rapid hydrogen evolution, droplets flew out of the solution, which were then transported by a stream of air into the flame. However, this procedure was not suited to quantitative analysis, serving for qualitative purposes only. A similar procedure, with some modification, was used by TÖRÖK for the quantitative determination of calcium. For some purposes, mechanical dispersion devices of various shapes were designed, without however proving to be good enough for use in quantitative analysis.

More important is the use of ultrasonics [209]. A disadvantage is the complicated nature of the equipment, and the fact that the efficiency of the dispersion process varies substantially with the decreasing liquid level.

Roughly the same applies to electrostatic dispersion devices [763], [737]. Their principle is illustrated in Fig. 8.11. Due to a very strong electrical field (5 to 15 kV) electrical forces form on the surface and disperse the liquid. Compared with other types, these devices have the advantage that concentration by impacts of droplets on the wall is very slight, because the electrostatic charge of the mist chamber and of the aerosol droplets is identical: thus, more than 50 % of the solution reaches the flame in the analysis of aqueous solutions. When using organic solvents, the authors succeeded in transporting almost all the original solution into the flame in the form of an aerosol. Besides the complicated character of the equipment, the main disadvantage of this procedure is the fact that the dielectric properties of the solution vary with the concentration of the element analysed: this disturbs the linear course of the respective calibration curves. Other salts and molecules having a dipole moment also interfere in the electrostatic dispersion process.

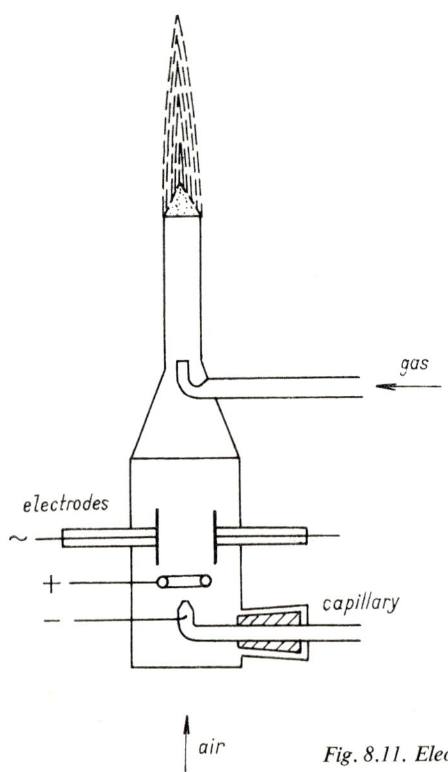

Fig. 8.11. Electrostatic dispersion device

8.3.3 Possibilities of working with solid samples

For the sake of completing the discussion, it should also be mentioned that facilities for introducing ground solid samples into the flame in a stream of air have also been described in the literature [152], [153], [294], [545]: other papers describe the use of a glycerine suspension [317], [508]. In the latter case, conventional direct-injection burners with a larger inner diameter of the suction capillary were used.

In some special cases, the sample has simply been held in the flame on a platinum spoon. To hasten the distillation of the element to be analysed, the sample was mixed with various reagents which react with it and form volatile compounds with the element required. For example, to determine lithium, rubidium, and caesium in silicates, the sample was mixed with the LAWRENCE–SMITH mixture (8 parts $CaCO_3$ + 1 part NH_4Cl [839]). To determine indium and thallium the sample was mixed with silver iodide, which forms iodides with the elements named [840].

8.4 The mist chamber

A discussion of dispersion devices would be incomplete without a description of the mist chambers, in which these devices are usually located. The main function of the chamber is to separate large aerosol drops. Chambers usually are made of glass. Their shape and size vary widely. Up to now, no single criterion is known for judging the suitability of these parameters, and the differences in size and shape mentioned are due rather to commercial reasons than to any attempt at better operation of the instrument.

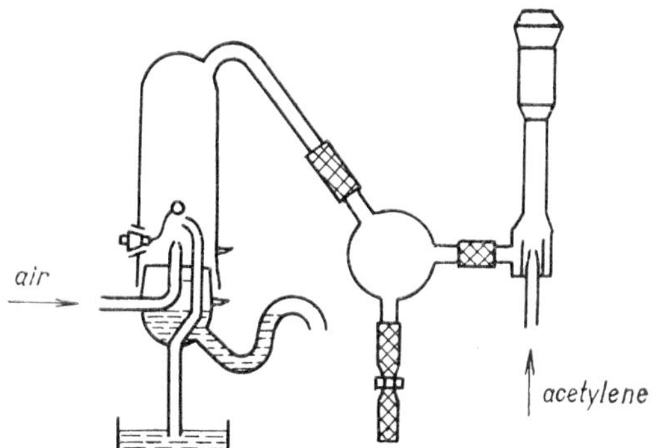

Fig. 8.12. Mist chamber with dispersion device according to RAUTERBERG *and* KNIPPENBERG

For example, the Model III Zeiss flame photometer has a relatively small mist chamber: it is cylindrical in shape and placed in a vertical position. In other instruments the chamber is spherical or conical in shape, and its volume is sometimes substantially larger, up to 2 litres.

Some types are shown in Fig. 8.12 to 8.16 [397, 565, 687, 767, 842]. Our experience shows [221] that a large volume of the mist chamber contributes to stabilisation of the aerosol inflow into the flame. The dispersion device is sometimes located at the side of the chamber, in other cases in its lower part: in some chambers the aerosol is introduced from above. Besides the different location of the dispersion device, there are also differences in the manner of removing the condensed aerosol. In some cases the condensate is allowed to flow away continuously, forming at the same time a siphon seal, in other cases it is only removed after the analysis is completed. The condensed solution is often re-used for analysis. This latter procedure, however, cannot be recommended fully because changes in the composition of the solution may occur as droplets condense on the chamber walls in the course of the analysis, which may bring into the solution analysed other elements and alter its composition, or they may dilute it, if water is involved.

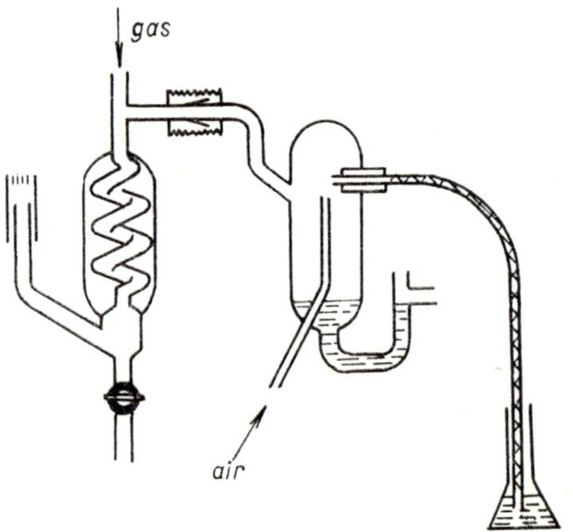

Fig. 8.13. Mist chamber proposed by SCHUHKNECHT

In the case of some instruments, where the mist chamber is small, a drop separator is included behind the chamber. With some types the air is already mixed with the fuel gas in the chamber, as seen in Fig. 8.16, but in the majority of instruments mixing takes place close to the burner, in the mixing chamber, or immediately in the body of the burner. Let us describe the differences between several of the most widely used instruments.

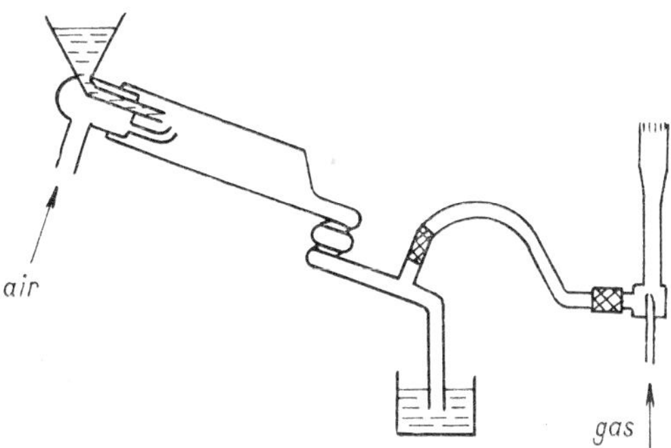

Fig. 8.14. Mist chamber and dispersion device belonging to the PERKIN—ELMER *flame photometer*

The Zeiss flame photometer is fitted with equipment described by RAUTERBERG and KNIPPENBERG, shown in Fig. 8.12. In this equipment the solution is brought from a vessel located outside the instrument. The aerosol flows vertically to the top and collides with the impact body. In the Unicam SP 900 flame photometer the dispersion device is located in a tangential position at the side of the mist chamber, the aerosol whirling in the chamber to avoid excess coagulation. A different shape is suggested by SCHUHKNECHT (Fig. 8.13). Another means of air inlet is used in equipment which forms part of the PERKIN—ELMER instrument (Fig. 8.14). The following figures illustrate mist chambers according to PINTA [574] (Fig. 8.15) and INGAMELLS [397] (Fig. 8.16).

Differing from direct-injection devices, the mist chamber type is relatively inefficient—only a small part of the solution analysed reaches the flame. For illustration, some values obtained with the Zeiss Model III flame photometer are presented in Table 8.4. With aqueous solutions this yield is not greater than 2 to 3 % in the case of most types of apparatus. Some authors, however, state that with suitable arrangement of the dispersion device in the mist chamber, the yield value may be increased up to 10 %.

In absorption flame photometry apparatus the above principles are maintained, and there is practically no difference between the two. The same equipment may be applied in absorption as well as in emission flame photometry.

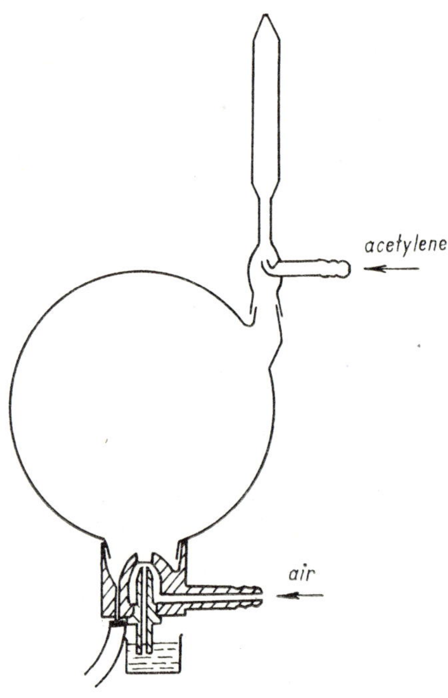

Fig. 8.15. Mist chamber with dispersion device according to PINTA

The solution to be analysed is usually placed in a small beaker, where the dispersion capillary is immersed into it. This arrangement is subject to the disadvantage that the level height of the liquid in the beaker varies in the course of the analysis, therefore the immersion depth of the capillary varies also and with it the rate of suction. However, there is the advantage that the surface of the solution is rather small, so that its evaporation is negligibly small in the course of the time needed for measurement.

The solution to be analysed is sometimes also poured into a little funnel, connected to the capillary in such a manner that gravitational energy

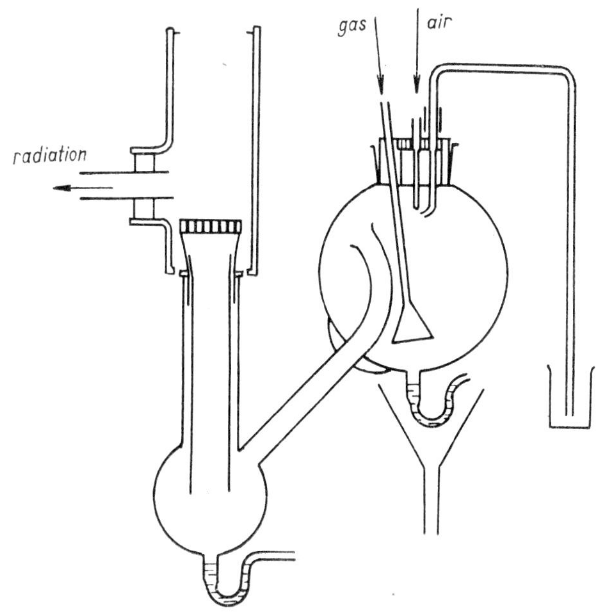

Fig. 8.16. Mist chamber with dispersion device according to INGAMELLS

Table 8.4. UTILISED PROPORTION OF THE DISPERSED SOLUTION
Zeiss Model III flame photometer

Solution composition	Proportion utilised, %
Water	1.7
20 % ethanol	4.4
30 % ethanol	6.9
40 % ethanol	8.1
50 % ethanol	8.4
70 % ethanol	9.5
90 % ethanol	10.5

aids the dispersion process. This principle is used exclusively in some instruments manufactured in the USA. Even with this type, however, the hydrostatic pressure is not maintained absolutely constant. Some authors therefore recommend that the rate of flow in the capillary be controlled by means of a syringe with regulated piston travel.

In the Zeiss Model III flame photometer the solution analysed is placed in a Petri dish, so that the liquid level remains practically constant throughout the analysis. Obviously, when highly volatile non-aqueous solvents are used there is substantial danger of evaporation, and the re-sulted changes must be kept in mind. The evaporation process may be slowed down by covering the dish with a lid.

8.5 Isolation of the required radiation

The wavelength range which includes the required radiation must be isolated from the remaining spectrum of the flame, which might interfere in the determination (the problem of imperfect separation of the radiation studied from the accompanying radiation, the so-called radiation interference, is discussed in detail in section 12.4). Separation may be achieved in principle in two ways: either with the use of light filters, or with monochromators. Both methods are conventionally employed in commercial instruments. Filters are used with simpler instruments which are cheap and do not require highly skilled operation and maintenance. It must be kept in mind, however, that the relatively great transmittance, which is the advantage of filters, is related to a lesser degree of monochromaticity of the transmitted radiation. The interval of wavelengths which are transmitted by the filter is wider than that of a monochromator in nearly all cases. Filters are well suited to the determination of those elements which have a very simple line spectrum (e.g. alkali metals), or to the measurement of molecular bands (e.g. alkaline earth elements).

Monochromators separate a narrower wavelength interval, so that the overall amount of radiation energy incident on the detector is smaller. Therefore, monochromators must be used in conjunction with a more sensitive radiation detector to compensate for this disadvantage. Obviously, monochromators are also more expensive and their operation and maintenance is more complicated and exacting. Their advantages include a greater degree of flexibility, the possibility of working with any wavelength desired and, therefore, a larger number of elements which can be determined with a single instrument.

8.5.1 Filters

Several types of filters are known. At present, however, interference filters are used practically exclusively in flame photometry. Coloured glass filters, which were used conventionally in former times, have a large degree of light transmittance (60 to 80 %), but the spectral region they cover is too wide, so that parasitic radiation interferes in the determination. In terms of quality, these filters are similar to those used for colorimetric purposes.

The most important types nowadays are interference filters. These are composed of two partly translucent reflecting layers, spaced at a distance equal to a low multiple of the selected wavelength. Multiple reflection takes place between the two layers, and the surrounding wavelengths are eliminated by interference of the transmitted rays. The secondary transmittance peaks are easily removed by combining the interference filter with light filters. The reflection layers are prepared by sputtering layers of metal (silver, aluminium) or several layers of translucent dielectric materials having alternately a large refractive index value (ZnS, Sb_2O_3) and a low one (MgF_2). In the first case, the interference filter achieves a peak transmittance value of 45 %, in the second case 80 to 90 %. The width of the transmitted band and the transmittance peak may be varied as required by combining several reflection layers [674]. The light transmittance characteristic of an interference filter is illustrated in Fig. 8.17.

A filter is characterised by the wavelength of its transmittance peak, λ_{max}, by the value of transmittance in the peak, usually denoted D, and by the width of the spectral region limited by one half the peak transmittance value. This latter is conventionally called the half-width. The ratio of the maximum to the minimum transmittance is usually greater than 300. The above three quantities suffice to characterise the form of the transmittance in the vicinity of the peak, which may be expressed by means of the relation

$$\tau = \frac{1}{1 + 4\left(\frac{\lambda - \lambda_{max}}{\Delta_{1/2}}\right)^2} \cdot \tau_{max}, \tag{8.1}$$

where τ is transmittance, and $\Delta_{1/2}$ is the half width.

The properties of filters used in Zeiss Model III flame photometers are listed in Table 8.5 as a specific example.

Interference filters must be kept in a dry place, because moisture may cause their characteristics to change. They are best kept in a hermetically sealed vessel (dessicator over a layer of silica gel).

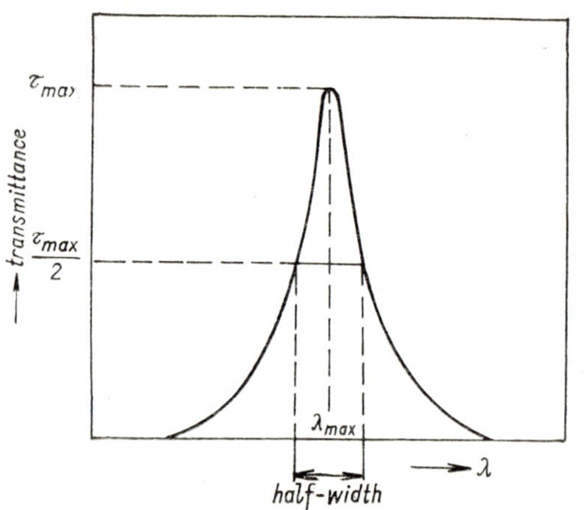

Fig. 8.17. Light transmittance characteristic of an interference filter

Table 8.5. CHARACTERISTICS OF SOME INTERFERENCE FILTERS MADE BY THE ZEISS CO. (GDR)

Notation	λ_{max} nm	τ_{max} %	$\Delta_{1/2}$ nm
Sr 46	459–463	25	10
Cu 51	508–512	25	10
Tl 54	533–537	25	10
Ba 55	552–556	20	10
Na 59	586–592	20	9
Ca 63	619–631	25	10
Li 67	668–674	25	12
K 77	762–774	40	20
Rb 79	780–793	25	15

One of the disadvantages of interference filters is the fact that the wavelength of the radiation studied is generally not exactly identical with the wavelength of the maximum transmittance of the filter, so that the entire transmittance of the filter cannot often be utilised. Care must also be taken when working with interference filters that the filter is placed in a position perpendicular to the direction of the incident rays. When the filter is inclined the transmittance peak shifts, since the distance between the two reflecting layers alters and with it the wavelength of the interfering rays.

8.5.2 Monochromators

Monochromators are optical instruments which serve to isolate a narrow spectral region. They consist of two parts: a dispersion system for spectral decomposition of the radiation studied, and an optical system which collimates and projects this selected radiation. Two basic types may be distinguished according to the dispersion system employed: prism, and grating-type monochromators. Their function is to be seen from Fig. 8.18. The radiation studied first passes through a so-called collimator, i.e. an objective lens at the centre of which the inlet slit of the monochromator is located. The collimator forms a parallel bundle of rays. Mirror-type objectives are frequently used, as these are not subject to chromatic aberration. The parallel beam of rays from the collimator then passes through the dispersion system and is decomposed. Another objective then forms images of the inlet slit of the radiation for radiation of various wavelengths. The outlet slit of the monochromator is located in its projection plane: when suitably adjusted it permits radiation of the selected wavelenth to be isolated. The best isolation of the required spectral interval is achieved when the widths of the inlet and outlet slits are equal. For this reason the two slits are usually controlled by a single mechanism.

In practice, of course, monochromators are usually designed according to the scheme shown in Fig. 8.18. They are frequently equipped with a LITTROW prism, one side of which is covered with an aluminium mirror, on which rays are reflected back into the collimation system, which at the same time focuses the image of the inlet slit onto the outlet slit. The wavelength of the transmitted radiation is not set by moving the outlet slit in the focal plane, but instead by rotating the dispersion element. The diagram of such a prism monochromator is seen in Fig. 8.19.

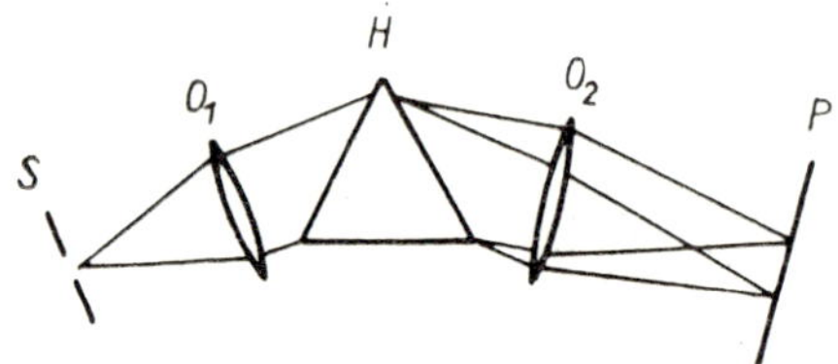

Fig. 8.18. Function of prism-type spectral instruments
S — entrance slit, O_1 — collimator objective, H — prism, O_2 — objective, P — focus area of objective O_2

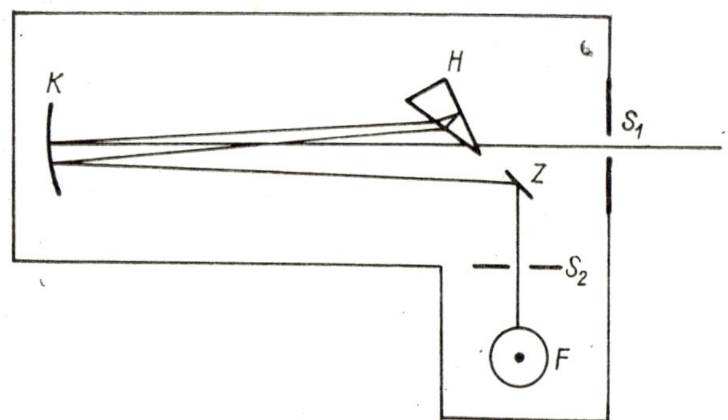

Fig. 8.19. Diagram of monochromator with LITTROW *prism*
S_1, S_2 — entrance and exit slits, K — collimation mirror, H — prism, Z — mirror, F — phototube

Spectral decomposition of light with a prism, as in Fig. 8.20. is based on the fact that the refractive index of the prism material differs for different wavelengths. Rays of lower wavelength are refracted on passage through the prism more than rays of a greater wavelength. This property, called dispersion, may differ with different materials. Of greater importance to the function of the prism, however, is angular dispersion, i.e. the angle separating two rays of a unit difference of wavelengths after passage through the prism. In the case of symmetrical passage through the prism, the following relation holds

$$\frac{d\varphi}{d\lambda} = \frac{2 \sin \dfrac{\alpha}{2}}{\sqrt{1 - n^2 \sin^2 \dfrac{\alpha}{2}}} \frac{dn}{d\lambda}, \tag{8.2}$$

where α is the refractive angle of the prism, φ is the angle of deviation of the transmitted ray, and n is the refractive index. Hence it follows that the difference in the deviation of two rays of slightly different wavelength will be the greater, the greater the refractive angle of the prism and the greater the dispersion of the prism material.

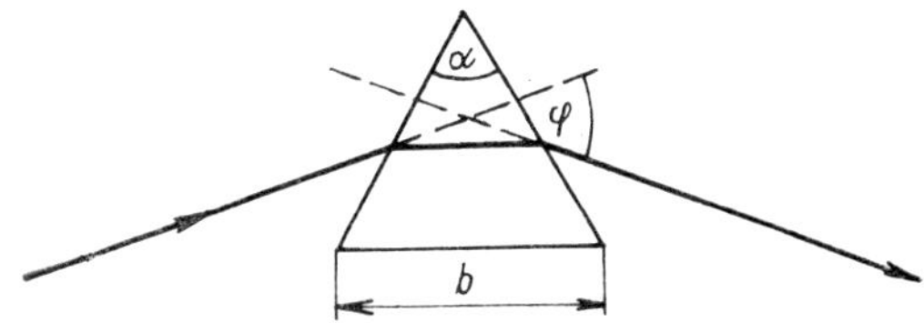

Fig. 8.20. Refraction of light by a prism

α — refractive angle of the prism, φ — angle of deviation of the ray, b — prism base

Prisms are mostly made of glass or quartz. Compared with quartz, glass has roughly a three-fold dispersion, i.e. it permits better resolution of close-spaced wavelengths. However, it absorbs radiation of a wavelength lower than about 370 nm, and for work in this region prisms must be made of quartz or other materials. The dispersion by quartz prisms in the visible region rapidly decreases with increasing wavelength so that closely-spaced wavelengths cannot be resolved.

Diffraction gratings have none of these disadvantages. They are made by cutting a large number of grooves in an aluminium mirror sputtered onto a carefully polished glass plate. Less perfect but cheaper gratings may be obtained by pressing the original into plastic and sputtering a mirror onto this replica.

On incidence of radiation onto the grating, reflection takes place on the mirror surfaces. The reflected rays continue in those directions for which the path difference of rays reflected on two neighbouring grooves corresponds to a whole multiple of the wavelength of the respective rays (see Fig. 8.21). This condition may be expressed by means of the relation

$$k\lambda = d\,(\sin\varphi + \sin\psi), \tag{8.3}$$

where d is the grating constant, and k is the so-called order of the spectrum. We find out easily by differentiating equation (8.3), that angular dispersion is independent of the wavelength of the radiation

$$\mathrm{d}\varphi/\mathrm{d}\lambda = k/d\,.\cos\varphi \tag{8.4}$$

It is likewise clear that angular dispersion is the greater, the higher the order of the spectrum and the smaller the grating constant, i.e. the larger the number of grooves per cm. Gratings for the ultraviolet and visible regions have usually 600 to 1200 grooves per mm. The first or second-order spectrum is usually used, since the intensity of the spectrum rapidly decreases with the rising order.

Monochromators are usually characterised by other properties. The resolution of the instrument

$$R = \frac{\lambda}{\Delta\lambda} \tag{8.5}$$

determines the minimum difference of wavelength of two bundles of rays which permits the images of the inlet slit to be resolved in the plane of the outlet slit. In the case of prism instruments, resolution is determined by the relation

$$R = b\frac{\mathrm{d}n}{\mathrm{d}\lambda}, \tag{8.6}$$

where b is the length of the prism base. With grating-type instruments, resolution is given by the product of the overall number of grooves on the grid and the order of the spectrum. The value of this theoretical resolution is usually substantially larger than the value of the actual resolution of the instruments. Another important characteristic of monochromators is the reciprocal linear dispersion. This depends on angular dispersion and on the focal length of the objective behind the dispersion

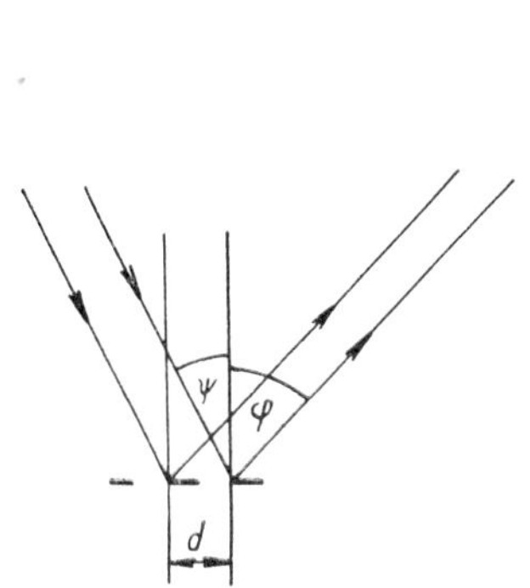

Fig. 8.21. Dispersion of light by a diffraction grating

d — grating constant, ψ — angle of incidence, φ — angle of reflection

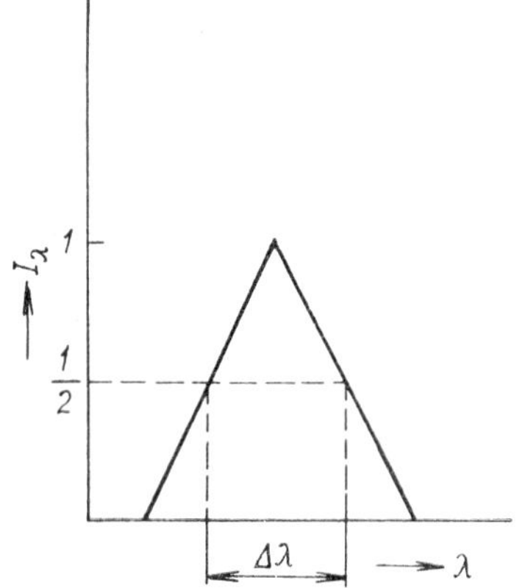

Fig. 8.22. Spectral interval transmitted by a monochromator with equal width of entrance and exit slits

Δλ — spectral slit width

system. When the focal plane is perpendicular to the optical axis, it holds that

$$d\lambda/ds = \frac{1}{f_2 \dfrac{d\varphi}{d\lambda}}, \tag{8.7}$$

where f_2 is the focal length of the objective. Since this gives the magnei tude of the wavelength interval $d\lambda$ corresponding to an element of length ds in the plane of the exit slit, it is possible to determine the width of the transmitted spectral interval from the known width of the monochromator slits. The slit width multiplied by the reciprocal linear dispersion is frequently called the spectral width of the slit. Its relation to the transmitted spectral interval, with equal widths of the inlet and outlet slits, is to be seen from Fig. 8.22. When the slits are narrowed down, the isolated spectral interval decreases down to the value of the so-called normal slit, which is determined by the following relation

$$A_{\text{norm}} = \frac{\lambda}{\dfrac{d\varphi}{d\lambda} d} = \frac{f_1 \dfrac{d\lambda}{ds} \lambda}{d}, \tag{8.8}$$

where f_1 is the focal length of the collimator, and d the effective diameter of the objective. On further narrowing of the slit below this value the amount of radiant energy transmitted rapidly decreases, while resolution is improved only slightly.

The amount of radiant energy transmitted by the monochromator obviously rises with the increasing slit width. However, cases must be distinguished where a line or a continuous spectrum are studied. In the first case the radiation flux passing through the outlet slit is determin d by the radiation flux passing through the inlet slit, multiplied by the transmittance coefficient of the optical system. This means that the radiation flux incident on the detector, and thus also the respective galvanometer deviation, increases in proportion to the slit width. In the case of a continuous spectrum, widening of the spectral interval of radiation transmitted by the monochromator will result as well as the abovementioned effect when the slit width is increased. The resulting radiation flux incident on the detector therefore rises with the second power of the slit width.

Hence follows the way in which the slit width of the monochromator must be chosen for different analyses. If the sample spectrum has a continuous background in the vicinity of the line studied, the slit must be made as narrow as possible, but not below the normal width value. In such a case, however, the transmitted radiation flux may be so small that the sensitivity of the detection system would have to be increased to such an extent that its own instability would become important, and the precision of the determination would be decreased. Thus, a compromise must be made. As long as the sample does not irradiate a continuous background in the flame, a larger slit width may be selected and the degree of amplification of the detection system may be lowered in proportion in order to achieve a sufficient measurement stability.

The optimum slit width is easily determined experimentally, by mesauring the deviations corresponding to the wavelength of the line peak I_{max} and the continuous background I_{back} as functions of the slit width. On plotting the ratio $\frac{I_{max} - I_{back}}{I_{back}}$, called the signal-to-noise ratio, against the slit width a curve as shown in Fig. 8.23 is obtained. The optimum slit width corresponds to the maximum of this curve.

Moreover, manufacturers usually state the optimum slit width in their catalogues for different wavelengths. The optimum values are generally of the order of tenths to hundredths of a millimetre.

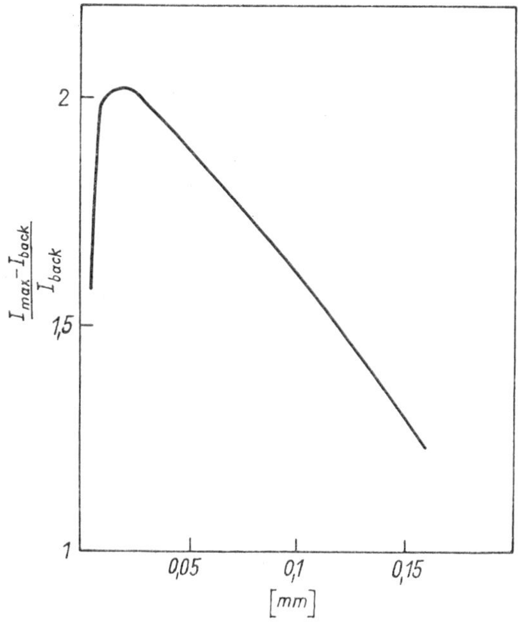

Fig. 8.23. Influence of slit width on the measured intensity ratio of the line and background (signal-to-noise ratio)

Flame photometer, Unicam Model SP 900; line measured, Sr 460,7 nm; I_{max} — intensity measured in the line peak, I_{back} — background intensity beside the line

8.5.3 Spectrographs

As mentioned already, a spectrograph may also be used to disperse the radiation emitted by the flame. The spectrograph is based on the same principle as the monochromator. Instead of the outlet slit, a photographic plate is located in the focal plane of an objective, where it simultaneously records all the radiation in a spectral range of the order of several hundred nanometres. This spectral range is usually limited by the design of the instrument or the spectral sensitivity of the photographic material employed.

Spectrographs have a substantially larger reciprocal linear dispersion than monochromators. In flame photometry, medium-dispersion spectrographs are usually chosen. Quantitative determination is done by measuring the degree of blackening of the individual spectral lines with the use of special photometers. For details, the reader is referred to special literature devoted to spectral analysis.

In respect of the advantages and disadvantages of this technique in flame photometry, the list of disadvantages may include the rather laborious procedure of measurement, and the fact that besides errors of flame photometric origin there are also the errors of the photographic plate, and the error of photometric evaluation: an advantage is the possibility of determining several elements at the same time. With many methods a photographic plate may be found to be a highly sensitive detecting device, as the incident radiation is integrated. The degree of blackening of the plate corresponds to the sum of radiation of a specified wavelength which fell onto the plate during the entire time of exposure, while photoelectric cells used in photometers only record the immediate value.

Lastly, a proposed quantometric apparatus, described by MARGOSHES and VALLEE [500] must also be mentioned. In principle this instrument does not differ from a spectrograph, but instead of a photographic plate it is fitted with photoelectric multipliers and the radiation of several selected wavelengths only is measured. This instrument permits several elements to be determined at the same time, with immediate evaluation of the results. At present the instrument is not yet being manufactured commercially, probably because of the great expense and complicated nature of the apparatus, which is not compensated by appropriate advantages in practical analysis.

8.6 Radiation detectors

8.6.1 Basic concepts

In flame photometry, the intensity of light radiation is measured almost exclusively by studying various electrical phenomena caused by the radiation. Sensitive devices used for this purpose are called detectors and, depending on the nature of the effect caused by the radiation, these may be classified in three groups:

1. Detectors based on the so-called external photoelectric effect

This principle underlies [374] vacuum photocells, gas-filled photocells, and photoelectric multipliers. When radiation falls onto the photocathode, electrons are liberated and transported into vacuum or into the gas phase. When a voltage is applied across the cathode and anode, a current is obtained (photoelectric current) which is proportional to the radiation flux incident on the photocathode.

2. Detectors based on the so-called internal photoelectric effect

In this system the electrons do not have enough energy to leave the cathode, i.e. they only increase the conductivity of the solid phase and permit the passage of a current when a voltage is applied. This effect is mainly manifested in semiconductors, having been observed for the first time in the case of selenium [713]. Photoresistors are based on this principle.

3. Detectors based on the barrier photoelectric effect

Barrier photocells are based on this effect. When a semiconductor joined to a metal electrode is illuminated, a potential difference is produced at the boundary between the two layers, and a current passes if the two are connected externally.

The most widely used detectors of this type continue to be selenium photocells [8], [466], although rapid development in the field of semiconductors has brought possibilities of the application of other semiconductive materials in recent years [205]. Detectors used in flame-photometric apparatus are photocells, photoelectric multipliers, and barrier photocells: photoresistors have found no application in this field up to now.

More complicated instruments are fitted with vacuum photocells, gas-filled photocells, or photoelectric multipliers, while less sophisticated instruments are usually equipped with selenium photocells.

8.6.2 Vacuum photocells

As mentioned already, vacuum photocells are based on the so-called external photoelectric effect, i.e. liberation of electrons from the surface of metals under the influence of light.

Liberation of electrons takes place when the energy of the light quantum is greater than the binding energy of the electron, i.e. the energy needed to liberate the electron. The energy of the light quantum depends on the wavelength of the incident radiation: with decreasing wavelength the energy of the light quantum increases. The binding energy of the electron depends on the nature of the metal from which the electrons are being emitted. The limit of the photoelectric effect, i.e. the greatest wavelength at which an electron can be liberated, differs with different metals. This means that an electron will be liberated from a given metal only if it is irradiated with light of a sufficiently low wavelength (sufficiently high wavenumber). With platinum, for example, the limit of the photoelectric effect corresponds to a wavelength of 280 nm, with potassium 600 nm and with caesium, 800 nm. Therefore a platinum photocell is suitable for measuring radiation in the ultraviolet spectral region, a caesium photocell is suitable for the visible and near infrared region.

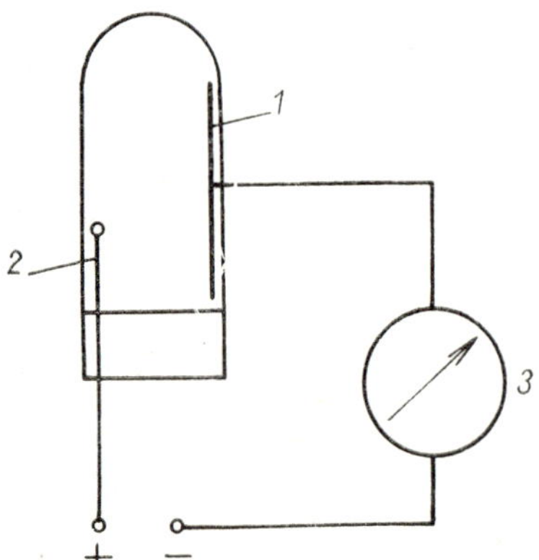

Fig. 8.24. Diagram of a phototube
1 — layer of alkali metal, 2 — anode, 3 — measuring instrument

Metals of the first group of the periodic system are usually used in making photocells, since their work function for electrons is lowest. The principle and function of photocells is illustrated in Fig. 8.24.

A photocell is an evacuated glass vessel with a light-sensitive layer of alkali metal applied to a silver-coated part. This layer, called the photocathode, is connected to a clamp on the outer surface of the photocell.

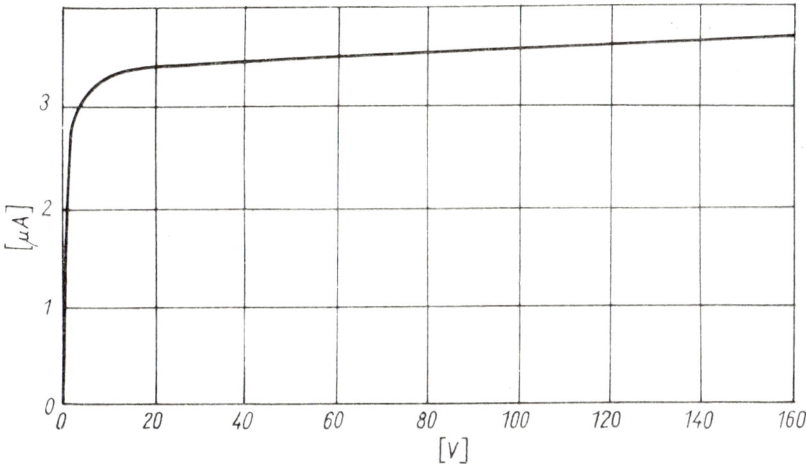

Fig. 8.25. Relation between current and external voltage with constant illumination, in the case of a vacuum phototube

The anode 2 frequently has the shape of a small disk which does not hinder the passage of light. The cathode is connected to the negative pole of a direct-current source (e.g. the anode of a battery) and the anode is connected to the positive pole. A sensitive measuring instrument 3 (e.g. a mirror galvanometer) connected into this circuit indicates no (or only a slight) current when the photocell is masked, since then the resistance of the photocell is great (10^{10} to $10^{14}\,\Omega$). When the sensitive layer is illuminated it starts to liberate electrons, the resistance of the photocell decreases, and a current starts to pass through the circuit, which fact is indicated by a deviation of the measuring instrument. The magnitude of the current depends on the external applied voltage and on the intensity of light incident on the sensitive layer. The relation between the current and applied voltage with constant illumination is illustrated in Fig. 8.25.

It will be seen that the maximum value is practically achieved at roughly 30 V. There is no purpose in using higher voltages, since they might cause changes in sensitivity.

The relation between current and light intensity is strictly linear in the case of vacuum photocells, which is the reason why photocells can be used for direct light intensity measurement. The sensitivity of commercial photocells varies from 5 to 80 μA/lm. From the principle of the external photoelectric effect it follows that the sensitivity of a photocell varies with the wavelength of the light measured. The wavelength range in which photocells can be used for measurement depends on the photocathode material. Nowadays a sufficient number of different photocell types is manufactured, so that the entire spectral range from ultraviolet through visible to near infrared can be covered by suitable combinations of photocell types. When photocells are used their operational data must be known, i.e. sensitivity, spectral characteristics, operational voltage, etc. These data usually are given by the manufacturer. The primary photoelectric current generally is too low. The circuit as illustrated in Fig. 8.24 generally is too insensitive for practical purposes. Therefore the photoelectric current commonly is amplified by means of electronic amplifiers.

8.6.3 Gas-filled photocells

These are based on the same principle (i.e. external photoelectric effect) as vacuum photocells, being however filled with inert gases (argon or neon). Electrons emitted by the cathode ionise the noble gas atoms, so that the photoelectric current is larger than with vacuum photocells, and thus the gas-filled photocells are more sensitive. The current magnitude depends, with constant illlumination, on the inert gas pressure in the photocell and on the external applied voltage. The applied voltage can be increased at will, as a discharge takes place if the voltage is too great.

Compared to vacuum photocells, gas-filled ones are four to ten times more sensitive. Their disadvantage is lower stability and a strong exponential relation between the photoelectric current and external voltage. Thus the practical application of gas-filled photocells is limited to the use of a highly stable external voltage source. These photocells are being used rather little in measuring techniques nowadays (687).

8.6.4 Photoelectric multipliers

The detectors used in modern commercial instruments with monochromators are almost exclusively photoelectric multipliers. This is the case, for example, with the instruments made by Perkin-Elmer, Unicam, Optics, Techtron, Opton, etc. The reason for the widespread use of these detectors is mainly their extreme sensitivity, which permits work in very narrow spectral intervals.

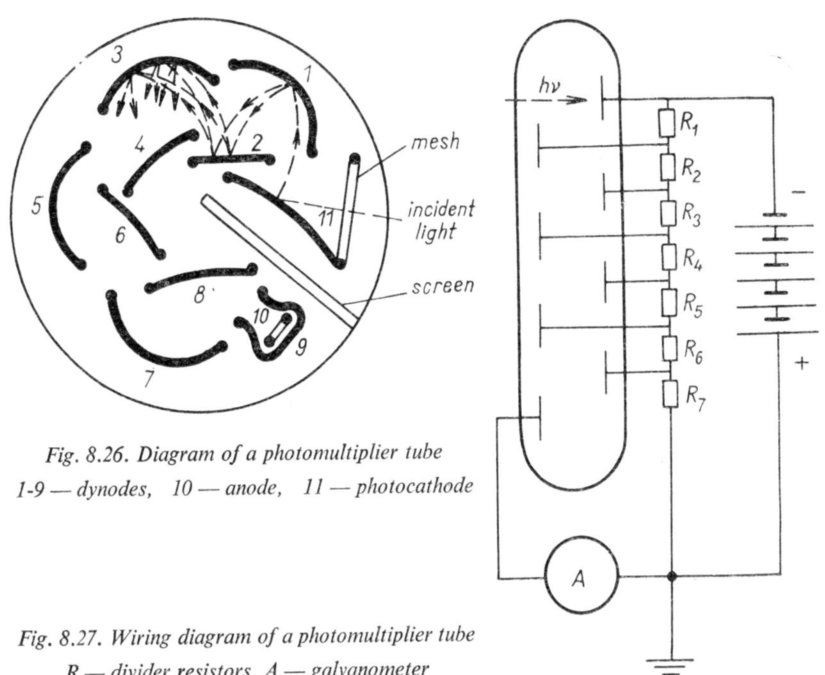

Fig. 8.26. Diagram of a photomultiplier tube
1-9 — dynodes, 10 — anode, 11 — photocathode

Fig. 8.27. Wiring diagram of a photomultiplier tube
R — divider resistors, A — galvanometer

Photoelectric multipliers, commonly called simply photomultipliers, are vacuum photocells with a built-in amplifier, the photoelectric current being amplified by secondary emission. Amplification factors of 10^6 and more can be achieved. Amplification takes place by liberation of secondary electrons from suitable emission dynodes, from which every primary electron can liberate several secondary ones. The principle of the photomultiplier is seen in Fig. 8.26.

A photoelectric multiplier consists of a cathode, anode and several dynodes, to which increasing potentials (with respect to the cathode) are applied with the aid of a voltage divider. The voltage between individual dynodes is usually 40 to 90 V. The wiring scheme is shown in Fig. 8.27. Primary electrons, liberated by a photon from the sensitive layer of the photocathode are attracted by and fall onto the first dynode, from which the impact liberates several more electrons. This process is repeated step-wise, the photoelectric current being amplified in the process. A measure of the yield of secondary electrons in one stage is the emission coefficient δ, defined as the ratio of secondary to primary electrons. The coefficient δ depends on the voltage per stage and on the dynode material. The resulting current thus may be expressed, in the case of photoelectric multipliers, by the relation

$$i = i_0 \delta^n, \tag{8.9}$$

where i_0 is the primary photoelectric current from the cathode, δ the emission coefficient of one stage, and n the number of stages. With $n = 10$ and $\delta = 4$, the resulting amplification will be $i/i_0 = 4^{10} \doteq 10^6$.

The resulting current of the photoelectric multiplier rises with the increasing voltage applied to the dynodes in practically an exponential manner. Hence it follows that the voltage source must be perfectly stable, which makes demands on the design of the instrument very exacting. Several batteries connected in series, or special stabilised sources may be used. Table 8.6 lists the spectral sensitivity peaks and limits of individual photocathode types. Photomultipliers with Sb–Cs photocathodes for the ultraviolet and visible, and Ag–O–Cs for the visible and near infrared region are used most often in flame photometry. Some other important properties of photoelectric multipliers are given in Table 8.7 [59].

The photo-current from the photoelectric multiplier can either be measured directly with a mirror galvanometer, capable of recording currents of about 10^{-9} A, or the current is fed to a precise resistance of the order of ten to hundred MΩ, where it is then measured with the aid of an electronic voltmeter. Frequently the current is first amplified before being measured. To remove short-term fluctuations of the flame, the signal from the photomultiplier may be converted to a mean value

Table 8.6. SPECTRAL SENSITIVITY OF PHOTOCATHODES

Type	Maximum spectral sensitivity	Long-wave limit nm
Ag–O–Cs	800	1200
	350	
Sb–Cs	400	650
Bi–Ag–O–Cs	450	800
Sb–Na–K–Cs	420	900

by charging a condenser with it for some specified time and then measuring the resulting charge. Details of the advantages of the individual wiring schemes will not be presented, as they lie outside the scope of this book.

8.6.5 Barrier photocells

The function of barrier photocells is based on the inner photoelectric effect. In this case, light energy is converted immediately to an electric current, without any external voltage source. The principle is shown in Fig. 8.28.

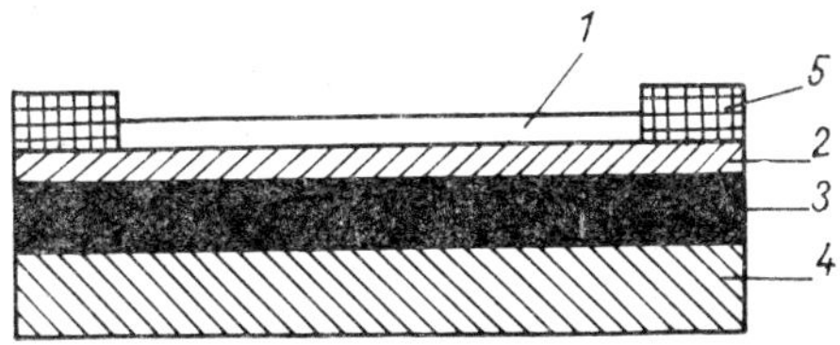

Fig. 8.28. Diagram of a barrier photocell

1 — protective layer, 2 — transparent metal layer, 3 — semiconductor layer, 4 — metal plate, 5 — contact

The semiconductor in this photocell may be cuprous oxide, thallium sulphide, etc. Nowadays, however, barrier photocells are made most often with selenium. The transparent metal layer is made of noble metals (platinum, gold) in thicknesses of roughly 4 microns. The metal base is usually a plate of iron. The spectral sensitivity of these photocells is compared with the sensitivity of the human eye in Fig. 8.29.

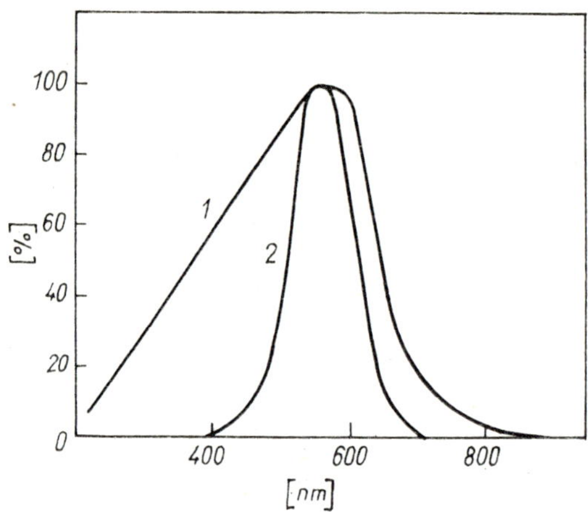

Fig. 8.29. Relative spectral sensitivity of a selenium photocell (1) and human eye (2)

The advantage of barrier photocells is the fact that their photoelectric current may be measured with the use of a very simple apparatus (a mirror galvanometer) and that they do not require a stabilised voltage source, so that the flame photometers fitted with barrier photocells can be very simple. Another advantage is the low price of the cells. In addition, the photocell delivers no current when not illuminated, i.e. there is no dark current which occurs with photoelectric cells and multipliers and which has to be compensated in the course of measurement. The photoelectric current formed in barrier photocells is not directly proportional to illumination. The form of the relationship between the photoelectric current and illumination depends, to a substantial degree, on the resistance of the external circuit. This relationship is illustrated in Fig. 8.30 [454] for various illumination intensity levels. Thus, to keep the relationship between the current and illumination at least roughly linear, low illumination intensities and a small external resistance must be used. In flame photometry the first demand is satis-

Table 8.7. PROPERTIES OF SOME PHOTOELECTRIC MULTIPLIERS

Type	Manufacturer	Cathode	Number of stages	Operational voltage	Amplification at operational voltage	Maximum dark current $\mu A/V$	Maximum output current μA	Note
6365	Du Mont	Sb—Cs	6	1000	$3 \cdot 10^3$	1/1000	500	
6364	Du Mont	Sb—Cs	10	1500	$1 \cdot 10^6$	0.05/1250	500	
6095	EMI	Bi—Ag—O—Cs	11	1500	$1 \cdot 10^7$	0.07/1500	1000	
6255B	EMI	Sb—Cs	13	1450	$5 \cdot 10^7$	0.5/1450	1000	for the UV region
6256B	EMI	Sb—Cs	13	1550	$5 \cdot 10^7$	0.03/1500	1000	for the UV region
9502B	EMI	Sb—Cs	13	1450	$5 \cdot 10^7$	0.015/1450	1000	
1P21	RCA	Sb—Cs	9	1000	$1 \cdot 10^6$	0.1/1000	100	
1P22	RCA	Bi—Ag—O—Cs	9	1000	$1 \cdot 10^5$	0.25/1000	100	
1P28	RCA	Sb—Cs	9	1000	$1 \cdot 10^6$	0.1/1000	100	for the UV region
5819	RCA	Sb—Cs	10	1250	$1 \cdot 10^6$	0.05/1250	100	convex photocathode
6342	RCA	Sb—Cs	10	1500	$1 \cdot 10^5$	0.05/1250	1000	
6810	RCA	Sb—Cs	14	2300	$1 \cdot 10^8$	0.75/2000	1000	time resolution 10^{-9} s
FEU17A	SSSR	Sb—Cs	13	1400	$5 \cdot 10^7$	0.3/1400	100	
FEU19M	SSSR	Sb—Cs	13	1900	$1 \cdot 10^7$	0.6/1900	100	
FEU20	SSSR	Sb—Cs	8	900	$1 \cdot 10^5$	0.008/900	—	
FEU22	SSSR	Ag—O—Cs	13	1400	$1 \cdot 10^5$	—	100	for the IR region
FEU35	SSSR	Sb—Cs	8	1400	$1 \cdot 10^5$	0.004/1400	—	
M12FD	Zeiss	Ag—O—Cs	12	1500	$1 \cdot 10^6$	0.5/1500	50	for the IR region
M12FS	Zeiss	Sb—Cs	12	1500	$1 \cdot 10^6$	0.1/1500	50	
M13S	Zeiss	Sb—Cs	13	1500	$1 \cdot 10^6$	0.5/1500	50	
M 120	Zeiss	Ag—O—Cs	12	1500	$5 \cdot 10^6$	0.1/1500	50	for the IR region

fied in practically all cases (since dilute solutions are used). The external resistance depends on the measuring instrument employed, which therefore should always have as low a resistance as possible.

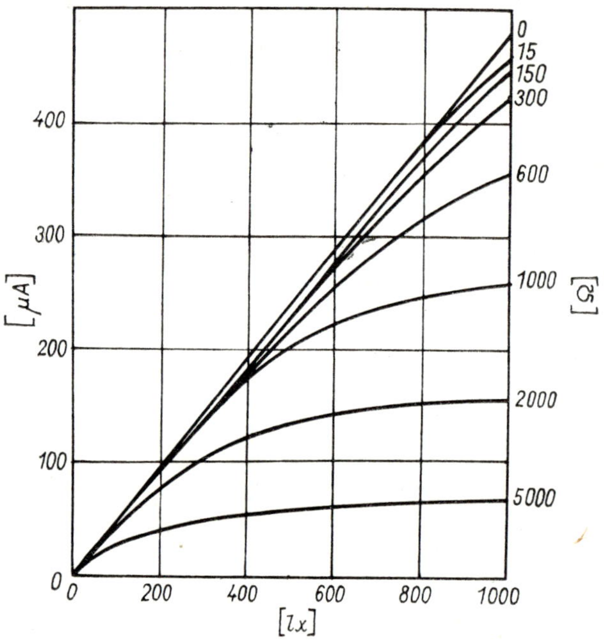

Fig. 8.30. Dependence of photoelectric current of a selenium photocell on illumination and external resistance

The non-linear relationship between photoelectric current and illumination in barrier photocells is in no way critical for flame photometric purposes. When a calibration curve is used for evaluation, a linear relationship is not an essential condition for accurate results to be obtained.

Another unfavourable property of barrier photocells is their "fatigue", which is manifested by a decrease of the photoelectric current with increased duration of illumination. This relationship is illustrated in Fig. 8.31 [384] for various illumination intensity values. It is seen that "fatigue" is less marked with low illumination intensities. It is moreover to be seen that, after an initial decrease, the current is stabilised on a certain value and then varies but little with time. Therefore, in flame photometry the instrument is allowed to "run in" at the beginning of the experiment, so that the measurement can take place in the stabilised part of the "fatigue" curve. Fatigue also depends on the wavelength of the radiation measured. Therefore, a photometer should be allowed to "run in" at the wavelength of the radiation to be measured.

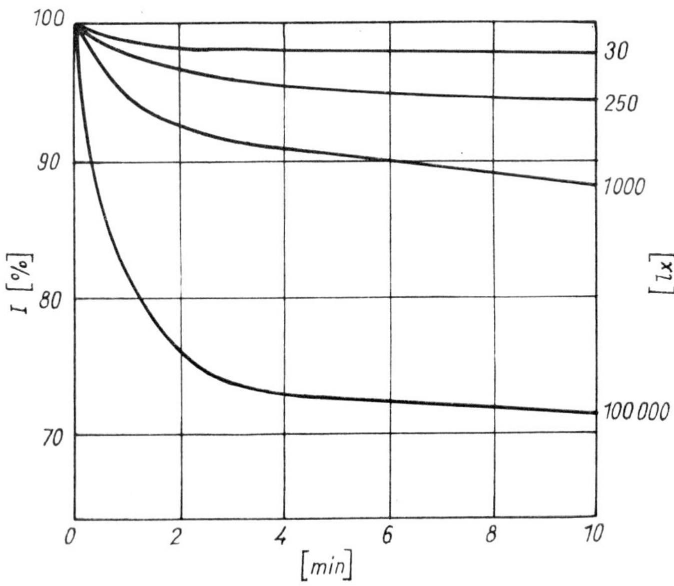

Fig. 8.31. Fatigue in a barrier photocell at various illumination intensities

9. Instruments used in flame photometry

9.1 Emission analysis instruments

Flame photometers may be classified from different points of view, most frequently according to the method used to isolate the required radiation. Thus, filter and monochromator instruments are distinguished. Within these groups, further classification may be based on the type of burner

Fig. 9.1. Model III Zeiss flame photometer

and flame employed, the number and quality of filters, or whether a prism or a grating is used for light dispersion.

Let us limit ourselves to some basic, frequently employed types. These will serve to illustrate principles common to most other instruments, though there are differences in detail.

9.1.1 The Zeiss Modell III flame photometer

One of the filter-type instruments frequently employed is the Zeiss Model III flame photometer. Its photograph is shown in Fig. 9.1, its diagram in Fig. 9.2. The instrument is designed for an acetylene-air or town gas-air burner. It is designed on the building-block principle, permitting the instrument to be modified or enlarged as required, and it is very simple in operation. It has given good service in many laboratories.

Air taken from a pressure bottle or a compressor is fed to the dispersion device through a rubber tube. The air pressure is measured with a manometer located on the measuring panel. The instrument is designed for an overpressure of 0.4 kg/cm^2(5.1 psi), though it is known from the literature that some authors have also used higher or lower pressures. The mist chamber forms a single unit together with the dispersion device: it was described in detail in section 8.4. With every dispersion device, two suction capillaries of an inner diameter of 0.4 mm and two of 0.6 mm are supplied. The consumption of the analysed solution is roughly 6 to 7 ml/min with an 0.4 mm capillary. The aerosol formed in the mist chamber is fed to a spherical drop separator and hence to a mixing chamber, where acetylene or town gas is introduced through a narrow capillary. The diameter of this capillary differs in mixing chambers for town gas and for acetylene: the two may not be interchanged. A tube connects the spherical part of the mixing chamber with the burner, which is divided on the inside into four sectors to ensure laminar gas flow.

Acetylene is introduced into the instrument from a pressure bottle, while town gas may be taken from the distribution piping. The pressure of these gases is measured with a water-filled manometer, also located

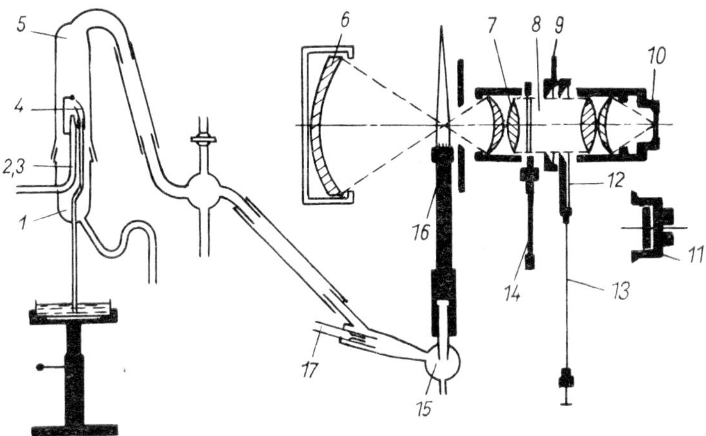

Fig. 9.2. Diagram of the Model III Zeiss flame photometer

1 — lower part of the mist chamber, 2 — air inlet, 3 — pressure nozzle, 4 — suction capillary, 5 — upper part of the mist chamber, 6 — hollow mirror, 7 — multi-lens condenser, 8 — iris diaphragm, 9 — diaphragm closure, 10 — ground-glass plate for adusting the burner, 11 — selenium barrier photocell, 12 — shutter, 13 — control trigger, 14 — disk with filters, 15 — mixer, 16 — burner, 17 — acetylene or town gas inlet

on the panel to facilitate pressure measurement in the course of the analysis. The acetylene pressure is adjusted experimentally (see section 10.2).

The burner for the acetylene-air flame is fitted on its upper part with a removable cap with holes 0.9 mm in diameter, preventing the flame from back-flashing into the burner, which might cause an explosion in the mixing vessel.

The flame burns between a concave mirror and a set of lenses which focus the image of the flame onto a selenium photocell. The mirror ensures optimum utilisation of the light flux. An iris diaphragm located between the flame and the condenser permits the light flux falling onto the photocell to be controlled. The current of the selenium photocell is measured with a galvanometer. A disc, into which four filters can be fitted, is located in front of the photocell. The required filter is placed each time in the path of the beam of light to be measured. The barrier photocell can be replaced by a ground glass screen marked with circles to facilitate adjustment of the burner.

The sample is placed in a PETRI dish, which guarantees a nearly constant depth of immersion of the suction capillary for a relatively long time. The PETRI dish is placed in a sample stand which is easily moved by a lever into the upper working position, or the lower position in which suction of the sample solution is interrupted.

9.1.2 The Unicam Model SP 900 flame photometer

This flame photometer is rather widely used. It is a single-purpose instrument fitted with a quartz prism and Type EMI 9529 B photomultiplier, sensitive from 250 to 850 nm. The instrument and its optical scheme are shown in Figs. 9.3 and 9.4. The sample is dispersed by a concentric device made of glass. The instrument is designed for an air overpressure of 0.8 to 1.0 kg/cm^2 (11.4 to 14.2 psi). The inner diameter of the suction capillary is 0.3 mm: it is made of a platinum (80 %) – iridium (20 %) alloy. A polyethylene capillary immersed in the sample vessel is fitted to the capillary. The consumption of the sample solution is 2.5 to 3 ml per minute. The aerosol formed is led to a MECKER burner where it is mixed with the fuel. An acetylene – air flame is usually used, although a burner cap for propane-butane is supplied with the instrument: the cap is easily replaced. The flame is ignited with an electric spark.

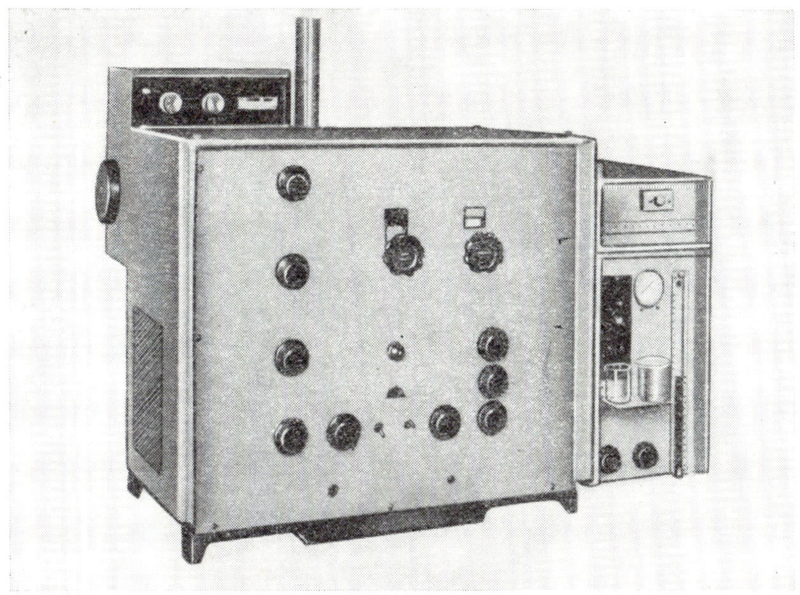

Fig. 9.3. Model SP 900 Unicam flame photometer
[by kind permission of Messrs. Pye Limited, Cambridge]

The optical system is formed by a prism and a set of mirrors arranged in such a way that the beam passes through the prism twice, which leads to better resolution. The prism has a refractive angle of 60° with a base length of 7 cm. A grey diaphragm may be placed in the ray path to permit the determination of large element concentrations in the solution analysed. Before falling onto the photomultiplier cathode the light beam is interrupted by a rotating chopper. The signal from the photomultiplier is then amplified by means of a synchronous a.c. amplifier, and after rectification it is read on a galvanometer scale. The instrument permits the zero position to be shifted in order to compensate for the background of spectra and changes in the degree of amplification, as well as adjustment of the time constant of the measuring system, and of the slit width. The entire control equipment including gas pressure control and indication is located on the front panel of the instrument, so that its operation is very convenient. The instrument may be fitted with a recording device which facilitates the resolution of radiation interferences. For atomic absorption the instrument may be amplified by the Model SP 900A equipment. Compared with atomic absorption adaptors used with other instruments (Hilger, Uvispek, Optica Milano Model CF 4), this instrument has the advantage of amplifying the chopped signal of the discharge lamp, so that the emission of the flame itself is eliminated.

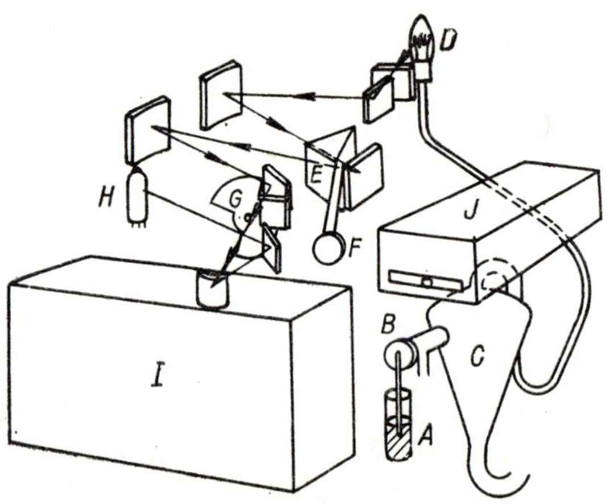

Fig. 9.4. Diagram of Model SP 900 Unicam flame spectrophotometer
A — analysed solution, B — dispersion device, C — mist chamber, D — burner, E — prism, F — control facilities, G — rotating sector, H — detector, I — instrument housing, J — galvanometer

9.1.3 Other instruments

An instrument used widely in Western countries is the BECKMAN spectrophotometer, several types being manufactured. The most widely known type DU is shown in Fig. 9.5. This instrument differs from the preceding mainly in the fact that it has a direct-injection burner. The manufacturer supplies several types of these burners, which differ in the inner diameter of the platinum suction capillary. The instrument permits acetylene-oxygen, hydrogen-oxygen, and other flames to be used.

One of the few grating-type instruments which must be mentioned is the Model CF 4 spectrophotometer made by the Italian firm Optica Milano (Fig. 9.6), which is supplied with equipment for emission flame photometry as well as for atomic absorption. A BECKMAN-type burner again is built into this instrument. As in the preceding case, photomultipliers are used for detection. For recording spectra, a fast recorder is supplied by the manufacturer.

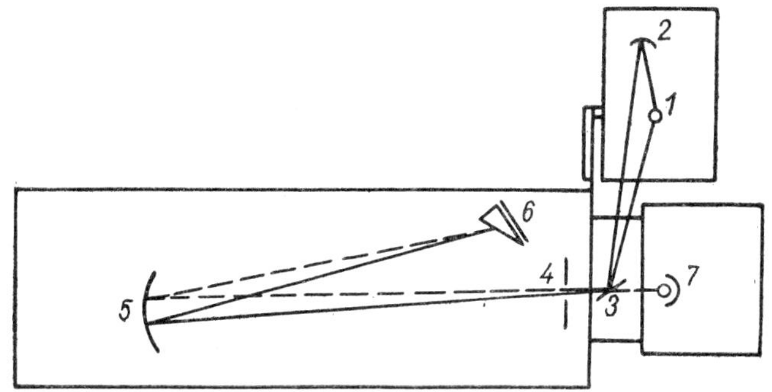

Fig. 9.5. Diagram of the BECKMAN *DU spectrophotometer*

1 — flame, 2 — mirror, 3 — planar mirror, 4 — slit, 5 — collimation mirror, 6 — prism, 7 — photomultiplier tube

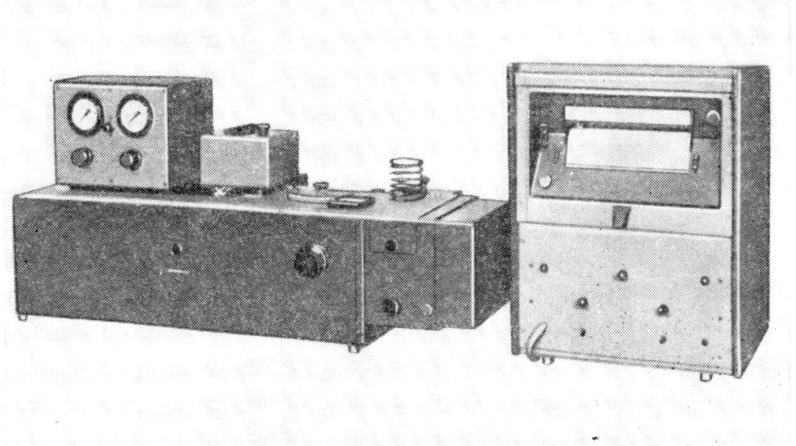

Fig. 9.6. Model CF 4 spectrophotometer by Optica Milano, with extension for flame photometric measurement

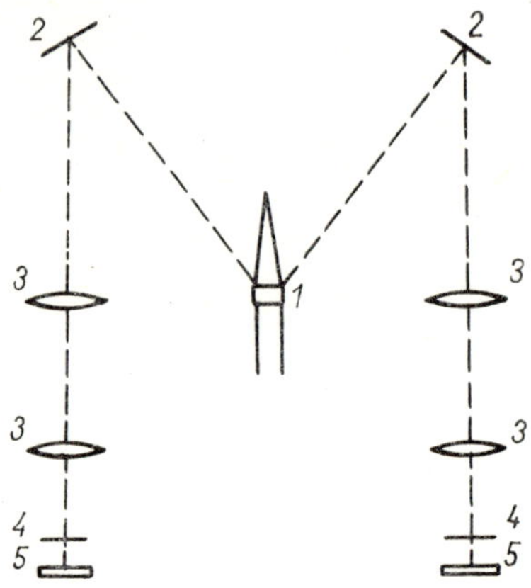

Fig. 9.7. Diagram of a double-beam instrument
1 — flame, 2 — mirrors, 3 — lenses, 4 — filters, 5 — radiation detectors

A special category includes instruments designed for simultaneous measurement at two wavelengths, e.g. the analytical line and the line of an internal standard (for details see section 10.7.5). The internal standard method is employed because comparison of the two values eliminates some errors of flame photometry, e.g. those related to the amount of sample introduced into the flame in a unit of time. The diagram of such an

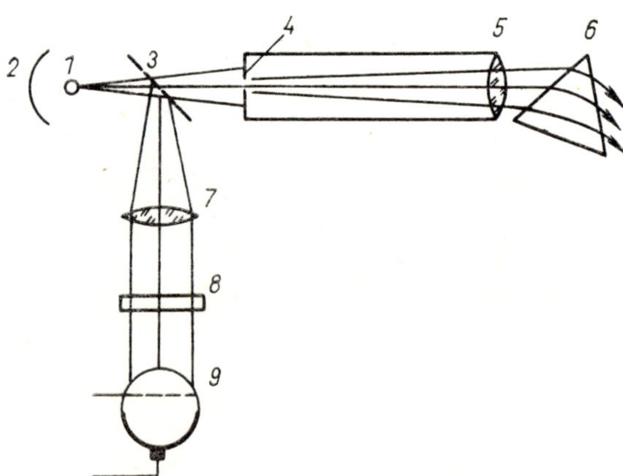

Fig. 9.8. Arrangement of the monochromator for the internal standard method
1 — flame, 2 — concave mirror, 3 — semi-transparent mirror, 4 — entrance slit of the monochromator, 5 — collimator lens, 6 — prism, 7 — collimator lens, 8 — filter, 9 — detector

instrument is shown in Fig. 9.7 [67]. Conventional spectrophotometers may also be adapted to internal standard measurement by means of adding a second detector, onto which the radiation of the internal standard falls after being separated by means of an interference filter, as shown in Fig. 9.8 [373].

9.2 Instruments for absorption flame photometry

Atomic absorption instruments include roughly the same parts as flame photometers, being in addition fitted with a source irradiating a line spectrum of the elements to be analysed. Although the main parts of the instruments are identical, they differ in some details which will be discussed in the following. The line spectrum sources usually are spectral lamps or, more frequently, hollow-cathode discharge lamps. Spectral lamps are manufactured for several easily volatile elements only, i.e. sodium, potassium, rubidium, caesium, thallium, zinc, cadmium and mercury. For this range of elements spectral lamps are supplied by the Osram Co., and by the Berliner Glühlampenwerk. In principle these are evacuated glass vessels with tungsten electrodes fused into the walls and containing a small amount of the element in the metallic state. The lamps are usually fed with alternating current at 220 V, being connected to a choke coil in series. When the current is switched on, a discharge is first formed between the main and the auxiliary electrode. When the vessel has been heated to the operational temperature, the volatile metal contained in the vessel evaporates and becomes a current carrier. At the same time the auxiliary electrode is switched off by a thermistor. These spectral lamps, however, do not irradiate sufficiently narrow lines, so that they are of rather limited utility. Therefore they are mainly used for alkali metals and mercury only.

Hollow-cathode discharge tubes are more generally applicable. Again, these are evacuated vessels filled with a noble gas, most often argon, with electrodes sealed into the vessel walls. The cathode has the shape of a hollow cylinder usually less than 1 cm in diameter. The anode may have any shape, and its position is not critical either. The outlet window is located opposite the cathode cavity. The gas pressure in the tube is in the range of 0.5 to 2 Torr. A hollow-cathode tube is shown diagrammatically in Fig. 9.9.

When a voltage of the order of 400 to 1000 V is applied to the electrodes, a discharge is formed. This discharge is concentrated near the cathode; the anode does not radiate. The discharge mechanism is the following: electrons, accelerated by the electrical field, collide with argon atoms, ionising them. The positive ions are attracted to the cathode, bombarding its surface and expelling metal atoms from it. This is called cathodic sputtering. These atoms arrive in a space where they collide with excited argon atoms (secondary/type collisions). Taking over their energy, they convert to the excited state and irradiate light quanta. Since a high concentration of sputtered atoms is maintained in the cathode cavity, the irradiation process also concentrates in this space. The spectra of hollow-cathode tubes are free from continuous background radiation and therefore their lines are very sharp, which fact is essential for high accuracy in the analysis.

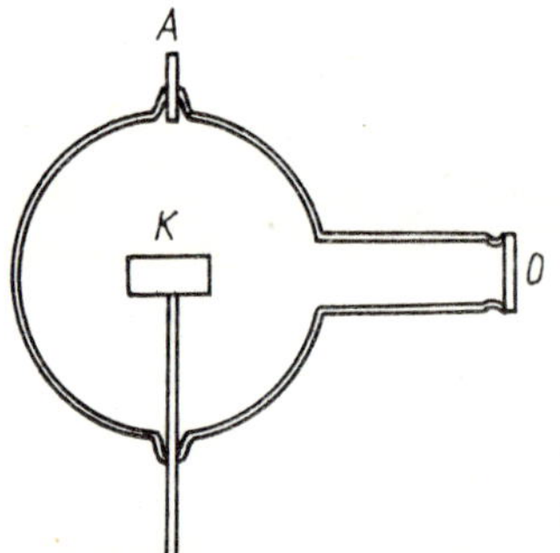

Fig. 9.9. Hollow-cathode discharge lamp
A — anode, K — cathode
The outlet window O is located on an extended tube to reduce its contamination by material sputtered by the cathode.

Discharge tubes are fed with a current of several milliamperes, occasionally several tens of milliamperes. It is recommended to choose the lowest possible current level which secures stable operation of the tube. Greater currents increase the radiation intensity, but at the same time the line width rises and the sensitivity of the determination decreases.

Discharge tubes are usually made in the shape of cylindrical or spherical vessels with volumes of at least 250 ml. With the decreasing volume of the discharge tube its lifetime is shortened, as this is determined by the sorption of noble gas atoms on the sputtered cathode material [420]. Opposite to the cathode cavity the vessel is elongated to form a tube ending in the outlet window. With elements whose resonance lines are located in the ultraviolet region, this window must be of quartz. The outlet window may be sealed with picein or epoxide resin.

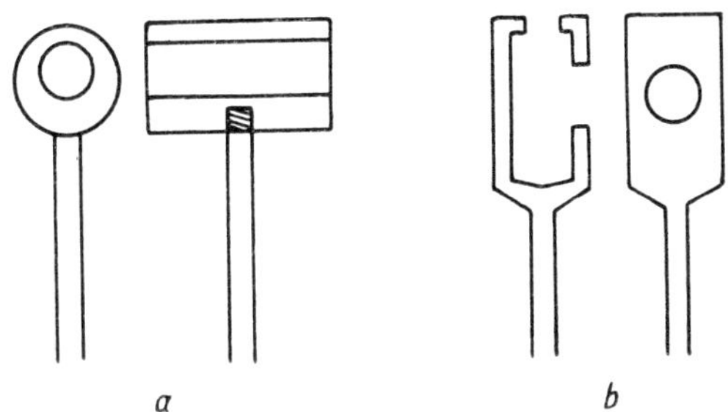

Fig. 9.10. Hollow cathodes
a — conventional shape, b — cup type for low-melting elements

The arrangement of the cathode depends on the physical properties of the element to be determined. With metals like copper, silver, magnesium, etc., the cathode may be made of the pure metal. With elements having a low melting point, e.g. zinc or lead, the cathode is made of some alloy. In the case of zinc, brass electrodes are generally used. Noble metals may be introduced into the cathode in the form of a wire net or foil. Elements of low melting points, the alloys of which are not generally available, may be introduced into the cathode in the form of a cup with a hole at one end to allow the radiation to pass through as seen in Fig. 9.10. Elements which are unstable in air, e.g. calcium, must be processed by special methods, or they may be introduced into the cathode in the form of one of their compounds. The body of the cathode is then made of a material which is sputtered very little, e.g. aluminium, duraluminium, steel, etc. If the cathode is to be replaceable, it must be sealed into a ground glass joint fitted into the tube vessel. This design is illustrated in Fig. 9.11. Obviously the ground joint must be lubricated with vacuum grease or sealed with picein wax.

When discharge tubes are made in the laboratory, it is advantageous to seal them with a vacuum stopcock and fit them with a ground glass joint for connection onto a vacuum filling apparatus [119]. In this way,

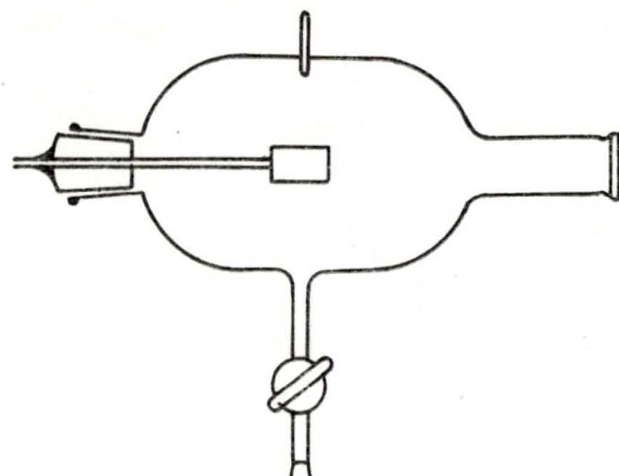

Fig. 9.11. Discharge lamp with exchangeable cathode and ground joint for connection to a vacuum apparatus

the discharge tubes can be renewed easily. A diagrammatic sketch of a vacuum filling apparatus is shown in Fig. 9.12. The apparatus consists of a rotary oil pump, mercury diffusion pump, two freeze traps, a closed mercury manometer, a McLeod manometer or other pressure gauge for the 0.1 to 5 Torr range and a noble gas reservoir, connected to the

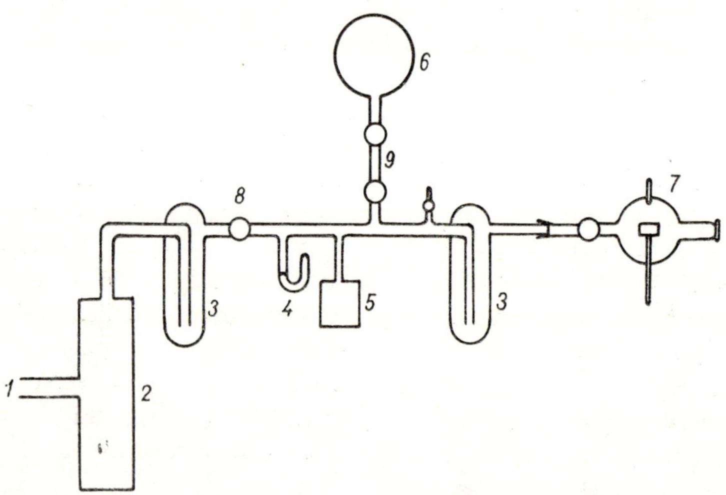

Fig. 9.12. Diagram of filling apparatus

1 — to rotary oil pump, 2 — diffusion pump, 3 — trap, 4 — manometer, 5 — McLeod gauge, 6 — gas reservoir, 7 — lamp, 8 — stopcock, 9 — dosage volume

apparatus through a dosage device. Argon is used most often. It should be spectrally pure, otherwise it must be purified before filling, which makes the apparatus more complicated. Some manufacturers supply argon in glass bottles 2.3 l in volume. On filling, the entire system is first evacuated to a pressure lower than 10^{-3} Torr and checked for leaks. Stopcock 8 (see Fig. 9.12) is then closed and a dose of argon is introduced. After this dose the pressure should rise to roughly 2 Torr. If a voltage is now applied to the electrodes, a discharge is formed, which is irregular at first until the the electrodes have been purified. When the pressure decreases, the spatial distribution of the radiation changes until it stabilises in the cathode cavity. On further lowering the pressure the smouldering discharge changes to a Townsend discharge, the radiation intensity decreases and the radiation fills the entire space between the electrodes. When the pressure is lowered even further the discharge is eventually quenched.

When the tube is filled for the first time, the cathode must be heated by increasing the current to 30 – 100 mA (depending on the material of which it is made). Obviously, care must be taken not to damage the cathode by the heat. After 5 to 10 minutes the lamp is re-evacuated, re-filled and heated. This process is repeated several times until all gases have been liberated from the cathode, which is best recognised by observing the red hydrogen line at 656 nm in the cathode radiation with a hand spectroscope. Some metals, e.g. iron or palladium, are very difficult to degas and the cathode must be heated for a substantially longer time in this case. The discharge tube is then set to the pressure at which the cathode cavity radiates most. It is then sealed or its stopcock is closed, and the tube is disconnected from the vacuum apparatus. The radiation stability of discharge tubes is excellent if stabilised current is employed. This is specially important with single-beam instruments, where it is recommended to stabilise the current to $\pm 0.25\,\%$. In practical operation, the lowest possible current is used. This decreases the extent of cathode sputtering and the lifetime of the tube is prolonged, and in addition the sensitivity of the atomic absorption analysis is improved.

Hollow-cathode discharge tubes are also manufactured commercially by some firms, e.g. Hilger, Westinghouse, Spectral Atomic, Quarzlampen G.m.b.H., Hitachi, etc.

Since the sensitivity of an atomic absorption determination depends on the thickness of the absorbing layer, burners 10 to 12 cm long are specially made for this purpose: the burners are oriented in the direction of the optical axis of the instrument. Slit-type as well as MECKER-type burners with separate holes may be used. The same applies to the size of these holes as was stated for burners in general.

Since in atomic absorption the atoms do not have to be excited, as it suffices to liberate them from the compounds in which they are bound, relatively cool flames are usually used, e.g. acetylene – air or propane – butane – air. The burners are made of cast iron, brass, or even glass. The upper plate is generally of nickel, rust-proof steel, tantalum, etc. It is usually water-cooled. To secure a regular distribution of the flame along the entire length of the burner, a plate is placed some 2 cm above the burner mouth. These burners are always designed for pre-mixed combustion gases, and therefore mist-chamber dispersion devices are used. In principle, however, direct-injection burners may also be employed.

Multiple passing by means of mirrors is sometimes used in order to elongate the path of the beam in the flame. Dispersion devices designed for absorption flame photometry do not differ in any way from those used in emission flame photometry. Generally they have wider suction capillaries as well as pressure nozzles, since the consumption of fuel gases is greater with long burners. Monochromators are usually used to isolate the required radiation. In the case of some elements which have a spectrum poor in lines, the line required may be isolated by means of coloured-glass or interference filters. The use of filters has been described for the determination of sodium, potassium, and mercury.

Radiation detection is similar to detection as practised in the emission method. When working with spectral lamps of relatively bright radiation selenium photocells are usually sufficient. With hollow-cathode discharge tubes, however, photoelectric multipliers must be used.

In radiation detection, either a continuous signal is measured, or a chopped one, the measurement being made directly with a narrow-band tuned millivoltmeter, or it is amplified by a tuned amplifier or one with synchronous detection, to be measured with a galvanometer after rectification. The first method with the continuous signal is employed generally with spectrophotometers fitted with an adaptor for atomic absorption, e.g. the Hilger Uvispek. A disadvantage of this method is the fact that with the increasing concentration of the element in the flame, its own radiation may interfere, causing the calibration graphs to become curves. This may be avoided in part by focussing the light source on the slit and de-focussing the flame image on the slit. In this way the proportion of radiation coming to the detector from the flame is decreased to a substantial extent.

With instruments designed specifically for atomic absorption modulated radiation from the light source is used. Modulation is achieved either by feeding the hollow-cathode discharge tube with network-frequency alternating current (the discharge tube radiates and quenches at a frequency of 50 or 60 c/s, i.e. it radiates only when the hollow electrode is the cathode), or by using two-way rectified current (in this case the light is modulated at a frequency of 100 or 120 c/s). Such a feed source and amplifier was described by Box and Walsh [88]. A rotating light chopper may also be used, and a corresponding greater frequency may be chosen. In this case the rotating chopper is placed between the discharge tube and the flame. An atomic absorption instrument of this kind is illustrated in Fig. 9.13.

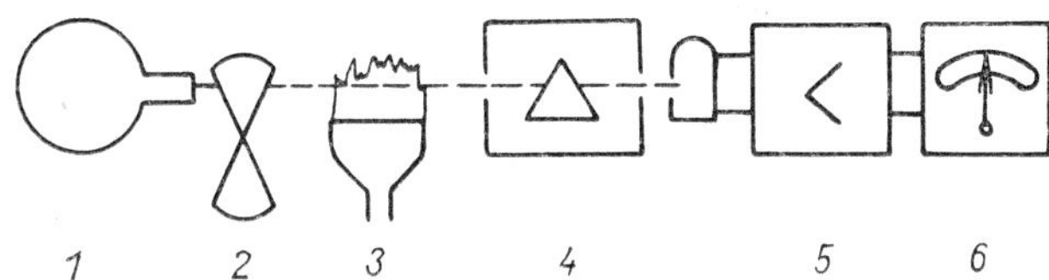

Fig. 9.13. Diagram of instrument for atomic absorption with modulation of light beam
1 — hollow-cathode discharge lamp, 2 — rotating light chopper, 3 — burner, 4 — monochromator, 5 — radiation detector and amplifier, 6 — measuring instrument

Up to now there are relatively few commercial atomic absorption instruments available. Among the single-purpose instruments designed specifically for atomic absorption we should mention the Model 303 instrument made by the Perkin – Elmer Co. (Fig. 9.14). This is a double-beam compensation-type instrument, in which all fluctuations of the light source, sensitivity of the photoelectric multiplier and amplification system are compensated automatically. The instrument permits the

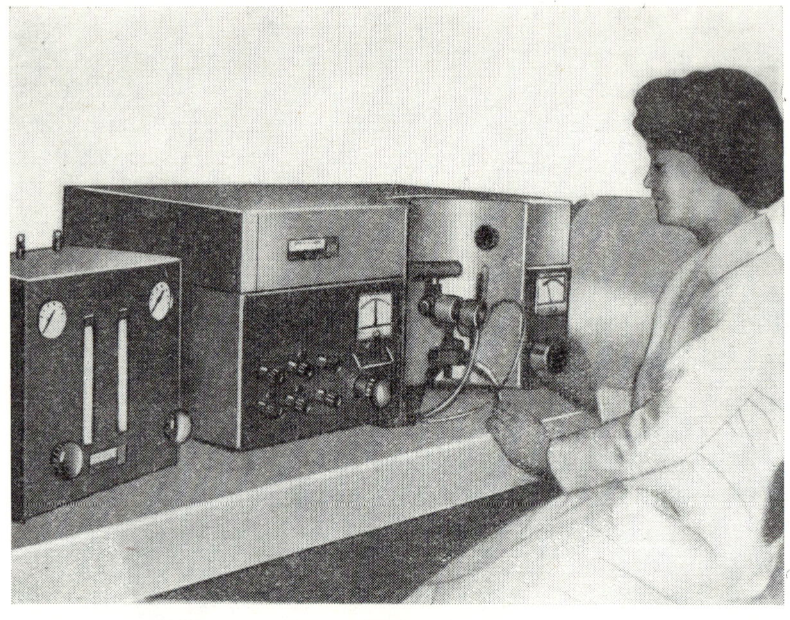

Fig. 9.14. Atomic absorption spectrophotometer, Perkin—Elmer Model 303 [by kind permission of Messrs. Perkin—Elmer Corp., Norwalk]

scale to be enlarged up to ten times, i.e. it permits an absorption value of 10 % to be measured over the entire scale. The optical diagram of this instrument is seen in Fig. 9.15.

Another instrument of this group is a product of the Research Control Instruments Co. This again is a compensation-type instrument, but of the single-beam class. It enables ten different hollow cathode tubes to be employed, converting from measurement of one element to another simply by moving a mirror. The Techtron Co. manufactures the Model AA4 and AA100 atomic absorption spectrophotometers. The absorption medium used in the instrument made by the Jarrell-Ash Company is a set of three direct-injection burners arranged in the optical axis.

Among the instruments manufactured in Europe, let us mention the Unicam Model SP90, Hilger Atomspek, Southern Analytical, and

Densatomic Optica Milano, many of which are designed in a way which allows them to be used for emission flame spectrophotometry also. All are fiftted with slit-type burners and allow the use of acetylene-air as well as dinitrogen oxide – acetylene flames. Other commercial instruments are simply adaptors made for use with different spectrophotometers. Nowadays, they are being built rather rarely, and single-purpose instruments are preferred in general. In specific cases, however, they are still justified, mainly from the economic point of view.

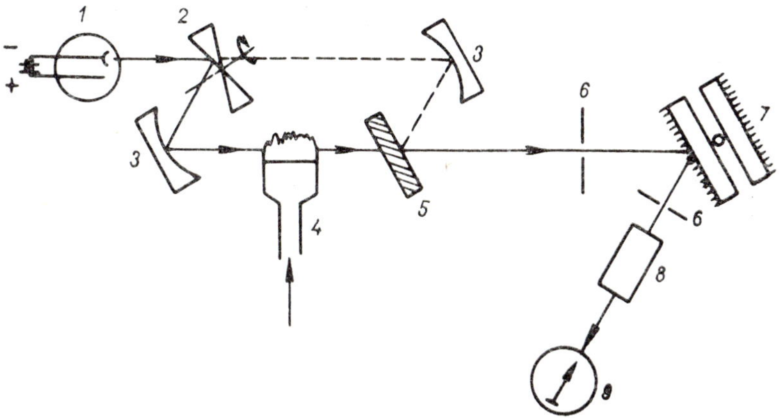

Fig. 9.15. Optical diagram of the Model 303 Perkin—Elmer spectrophotometer
1 — hollow-cathode discharge lamp, 2 — rotating mirror, 3 — concave mirrors, 4 — burner, 5 — semitransparent layer, 6 — slits, 7 — gratings, 8 — detector, 9 — indicator

The most widely used one is doubtless the adaptor for the Hilger Uvispek. It consists of a burner with cover, discharge tube with feed source, and dispersion device. It is a single-beam compensation instrument. Since it measures a non-modulated signal, an image of the cathode cavity is projected onto the slit to eliminate the radiation of the flame, while the flame itself is defocussed. Thus its radiation falling onto the detector is weakened substantially.

Many analysts prefer to construct their own instrument. As an example, let us mention a simple instrument composed of readily available parts. Its diagram is shown in Fig. 9.16. The source for the hollow-cathode discharge tube is a Type BS 275 Tesla instrument which gives a maximum current of 70 mA at 700 V. The value of the resistor connected in series with the discharge tube is 5 kΩ. For the dispersion device and burner, parts of the Zeiss Model III flame photometer may be used, the monochromator may be for example a Zeiss Model SPM 1.

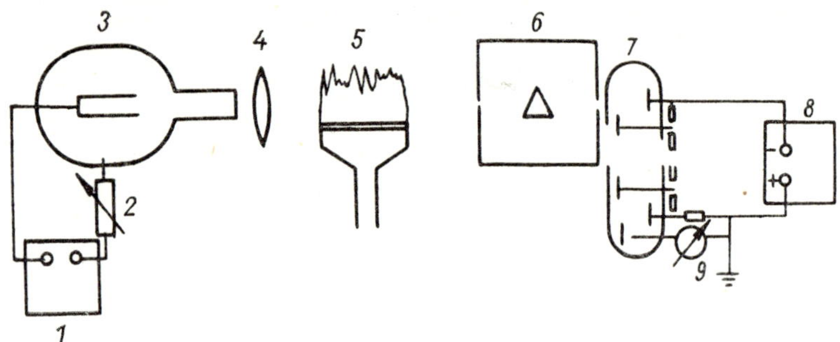

Fig. 9.16. Diagram of a simple atomic absorption instrument

1 — stabilized direct current source, 2 — variable resistance, 3 — hollow-cathode discharge lamp, 4 — lens/projecting the cathode onto the slit, 5 — flame, 6 — monochromator, 7 — photomultiplier with voltage divider, 8 — high voltage source, 9 — mirror galvanometer

A type FEU 18 A photoelectric multiplier fed from six anode batteries is fitted to the monochromator exit slit. The current from the photomultiplier is measured with a scale galvanometer which again belongs to the Zeiss flame photometer. In this arrangement the instrument is suitable for analysing elements whose resonance lines lie in the ultraviolet spectral region (14.3.3) or for which spectral lamps may be used instead of the hollow-cathode tube (14.3.4).

10. The technique of work with a flame photometer

10.1 Laboratory equipment

The instrument should be used in an evenly lit, well-ventilated room, free as far as possible from dust.

The instrument should be placed on a bench which is protected from shock. It is advantageous to have access to the flame photometer from all sides, and to have enough space on the bench for the solutions to be analysed. If the instrument consists of several parts, like for example the Zeiss Model III, it may be arranged on the bench in the manner illustrated in Fig. 9.1. For reasons of safety it is recommended to install a suitable ventilation device above the flame.

Pressure bottles containing gases must be placed as far as possible from the flame. In most countries, government standards prescribe rules for handling and storage of pressure bottles. At least fire extinguishers must be available in the laboratory.

10.2 Installation and adjustment of the flame photometer, illustrated with reference to the Zeiss Model III

Procedures for installing instruments and putting them into operation obviously differ with different instruments. However, many operations are common to most of them, and we shall describe some of the more important aspects using the Zeiss Model III flame photometer as a typical example. The design of the instrument is seen in Fig. 9.2, and the instrument has been described in detail in section 9.1.1.

The component parts are joined by rubber tubes. Before starting the adjustment of the dispersion device, the pressure bottles with air and acetylene must be connected to the instrument. Fitting the reduction valve to the bottle, opening and adjusting it is not difficult: yet, for operators who have had no previous experience it is better to do this at first in the presence of an experienced operator. When fitting reduction valves to pressure bottles, special care must be taken that they seal well and that no gas leaks into the laboratory atmosphere. The reduction valve supplied by the Zeiss Company is shown in Fig. 10.1. The gas inflow is controlled by means of a screw with a cross nut located on the lower part of the reduction valve or by the screw with star-shaped nut at the side. Leaks may not be searched for with an open flame, as this may cause an explosion. If it is required to test for the existence of small leaks, a soap solution may be used to good effect.

The first major operation to be carried out is adjustment of the dispersion device. This is done after the dispersion device has been removed from the mist chamber. The mouth of the suction capillary is submerged in water, and the optimum setting is determined at an air pressure of

Fig. 10.1. Reduction valve

some 0.2 to 0.3 kg/cm^2 (2.8 to 4.2 psi) by rotating the dispersion nozzle. The character of the mist formed is best observed against a black background, the direction of flow of the aerosol being perpendicular as far as possible. When the optimum position has been found, the air inlet is closed, and the dispersion device is fitted into the mist chamber. The support bearing the Petri dish is placed below the suction tube and the position of the mist chamber is adjusted in such a way that the lower end of the suction tube is only 0.5 mm above the bottom of the Petri dish when this is in the upper position.

The Zeiss flame photometer is fitted witth a selenium photocell, which must be protected from light until it is placed in its position in the instrument. The photoelectric cell is connected to the galvanometer by means of two-single-strand cables roughly 1 m long. Unless the laboratory or the galvanometer itself is sufficiently electrically screened, substantial interfering currents may occur, which however can be eliminated by grounding the instrument. The grounding wire is connected to the rear wall of the galvanometer by means of screws.

Before the selenium photocell is fitted to the instrument, the burner must be adjusted in operation. The prescribed procedure of igniting the flame must be adhered to exactly. First the pressure air inlet is opened and the normal operating overpressure is set to 0.4 kg/cm^2 (5.7 psi). At this pressure distilled water is dispersed. The acetylene inlet should not be opened before the liquid drops regularly out of the water seal of the mist chamber into a vessel placed below. The acetylene pressure is then set so as to achieve a difference of levels of the water column of roughly 60 to 70 mm. Before igniting the flame, check whether the outflow tubes of the separator and mixing chamber are closed. When the acetylene and air have streamed through for some 10 seconds, the aerosol and gas mixture in the burner is ignited. When acetylene is being purified by bubbling through concentrated sulphuric acid, and fresh acid is in the gas washing bottle, it takes some time for the acid to become saturated with acetylene: in this case the pressure will rise rather slowly, as seen on the water manometer. If this is the case, only a very small amount of acetylene will enter the burner for some two or three minutes, so that the mixture cannot be ignited. After several minutes the sulphuric acid is saturated and the gas mixture can be ignited. For reasons of safety a burning match should be held over the burner. This is the best way of finding out when the mixture is combustible, and escape of acetylene into the laboratory atmosphere is avoided.

The gas mixture should burn with a regular, erect flame practically with no noise, and the flame tip should move only very little. The blue-green cone of the flame above the burner head should be 2 to 3 mm high. If the gases contain excess acetylene, the blue cone becomes higher and a weakly luminous layer forms around it: with an increasing acetylene excess this layer converts to a bright and sometimes strongly sooty flame. With an excess of air a greatly irregular, mostly very noisy flame forms. This flame is extinguished very easily; its blue-green cone is very low and has a tendency to separate from the burner. In both cases the acetylene pressure must be adjusted.

The hottest part of the flame is immediately above the tip of the blue-green cone, and this is the part of the flame which usually is to be projected onto the sensitive layer of the photoelectric cell by means of the set of lenses and the concave mirror. The mirror permits the radiation which is emitted by the flame in the opposite direction to be turned back toward the photocell. Thus the direct as well as the reverse image of the flame are projected onto the photocell. A ground-glass screen, supplied with the instrument and usually fixed within it, serves to adjust the burner. When adjustment of the burner has been completed its housing is closed and the ground glass screen is replaced by the photocell.

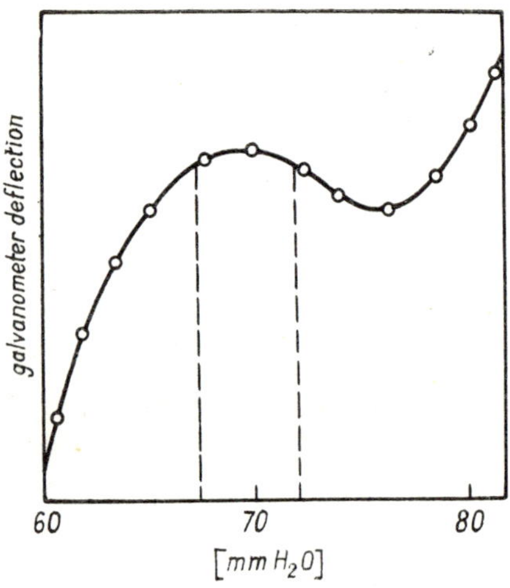

Fig. 10.2. Determining the optimum pressure of acetylene

The air pressure is determined by the design of the instrument, and it should not be increased: it should be kept constant at a value of 0.4 kg/cm^2 (5.7 psi). The most favourable acetylene pressure is achieved when working at the point of inversion of acetylene with the specified air pressure. Let us explain the procedure of determining this parameter with the aid of an example and an illustration. A medium-concentration standard solution is dispersed at a constant pressure. We start by measuring at an acetylene pressure of about 40 mm and increase this value gradually with the aid of the reduction valve on the acetylene pressure bottle. The rise should be approximately 5 mm water column. With every acetylene pressure value we determined the respective galvanometer deviation. The measurement is continued until an intensely luminous layer is visible in the flame above the blue cones. We now plot the galvanometer deflection versus the acetylene pressure. This gives a curve roughly corresponding to a third-order function (see Fig. 10.2).

The most favourable pressure corresponds to the peak of this curve: in the given case it is 70 mm water. This pressure represents a constant of the instrument which must be determined before any practical analyses can be done. However, the constant varies if any parts of the dispersion device, e.g. the suction capillary of the mixing chamber, burner, etc., are replaced. In such a case the constant must be re-determined. An experienced analyst is usually able to determine the optimum operating conditions directly from the appearance of the flame rather accurately, so that he has to determine the "point of inversion" rather rarely.

The pressure of both gases must be followed on the measuring panel in the course of the analysis. If the pressure varies, it must be adjusted to the original state. With a purposeful arrangement of the apparatus this checking and the measurement can be done by a single operator.

Work with the town gas – air flame

Most flame photometers may also be used for other gas mixtures. Usually equipment for the use of town gas is supplied together with the equipment for the acetylene – air flame. Since the pressure of town gas, which is taken from the distribution piping, is substantially lower than that of acetylene taken from a pressure bottle, the resistance of the instrument to the passage of this gas must be decreased in an appropriate way in order to obtain a gas mixture of suitable composition while maintaining an equal air pressure (0.4 kg/cm^2 or 5.7. psi). The adjustment is mainly achieved by fitting a different mixing chamber into the instrument, with a wider gas inlet capillary. Moreover, the dust filter which is mounted on the measuring panel must be disconnected from the town-gas inlet tubing and replaced with a simple glass tube. The gas pressure is then set to the optimum value in the way already described For reasons of safety this apparatus cannot be used for acetylene.

The disadvantage of the use of town gas is a tendency for the pressure to fluctuate. To regulate the pressure at least partly, the equipment described in section 8.1.6 may be employed. It is also better to use a flat burner, which is usually supplied together with the other accessories. When a MECKER burner is used for town gas, the blue cones of the flame usually are too high and their height moreover is not constant due to the oscillating town gas pressure, so that optimum adjustment of the burner is rather difficult. The procedure of opening the gas inlet and closing it, as described for acetylene, applies to town gas also. First the air inlet is opened, then that of town gas: on closing, the gas flow is shut off first, then the air inlet. The same rules as already described apply for adjustment of the burner.

10.3 Measurement with a flame photometer

10.3.1 Simple filter instruments, illustrated by the Zeiss Model III

Certain principles must be maintained in order to obtain accurate and precise results with a flame photometer. Firstly the instrument should be allowed to run for some time before the measurement is made, for the following reasons.

1. Although in principle measurements are carried out at room temperature, a decrease of temperature takes place on dispersion of the solution in the mist chamber, where a constant temperature is achieved only after several minutes. For this reason, measurements should not be interrupted for periods of more than one minute in order not to disturb the temperature equilibrium between the mist chamber and the room. If the measurement has to be interrupted, pure water should be dispersed during this time. When organic solvents are used, the respective mixture should be dispersed during the period of interruption.

2. Electrical and thermal effects due to the currents passing through the apparatus arise in the electrical parts of the instrument and in the galvanometer: these achieve a state of equilibrium after some twenty minutes.

3. It is essential for the "fatigue" of the photoelectric cell or other detector to stabilise.

The measurement is generally started with dispersion of the standard solution of maximum concentration. It happens very frequently that the radiation intensity is too great, so that the light transmittance of the condenser system must be decreased by partly closing the diaphragm aperture. In other cases, especially with very low intensities, it may be necessary to open the diaphragm completely in order to obtain a sufficiently large galvanometer deflection. Obviously the diaphragm must be opened to the same extent throughout the measurement, being set to the optimum position before the analysis is started. It is likewise important to set the zero value correctly. This is determined by the galvanometer deflection for the zero solution, i.e. a standard which does not contain the element to be analysed. This value usually differs from zero, and it must be taken into account when constructing the calibration curve. When the galvanometer has been correctly set by means of the standard solution of maximum concentration, other standard solutions are dispersed one after the other, starting with the lowest concentration, and the respective galvanometer deflections are noted in the form of a table. During the analysis the pressure of the two gases must be checked on the measuring panel. Measurement of the sample solutions is started immediately after the measurement of the stantard solutions has been completed. Every measurement must be repeated several times and after every five measurements the zero value of the galvanometer must be checked for possible variations. After the end of the measurement, distilled water must be dispersed for several minutes to clean the apparatus. After this the gas inlet may be closed. Again, a certain procedure must be maintained, as when opening the gas inlet. First the valve on the acetylene pressure bottle is closed. When the flame has burned down and all instruments on the measuring panel indicate zero values, the screws on the reduction valve are closed. Finally the air inlet is closed also. When this has been done, the entire apparatus is out of action.

10.3.2 Instruments with monochromators

Instruments fitted with a monochromator require a somewhat different technique of operation. The instrument itself is rather more complicated, as it includes a relatively large electronic block. Many parts of the instrument are of the compensation type, and therefore the sensitivity adjustment is done in a different manner from the preceding. The basic rules of operation, however, are not complicated. Firstly, as with other flame photometers, the adjustment of the burner position must be carried out and the correct pressure of the fuel gases must be determined. When the two values are known (they are generally stated in the manufacturer's instructions) the burner is ignited and the standard of maximum concentration is dispersed. Then the wavelength at which the analysis is to be carried out is selected by means of the conventional facilities. The accurate setting position must be found experimentally, as the wavelength scale on the instrument does not usually correspond exactly to the wavelength transmitted. The procedure involves switching on the detection equipment, waiting for it to stabilise, and then slowly rotating the wavelength selector while following the indicator at the same time.

It is then easily recognised when the intensity of the emitted radiation is maximum corresponding to the maximum indicator deflection. Should the difference between the wavelength read on the indicator scale and the true wavelength be too great, it can be decreased in the case of some instruments by adjusting the wavelength selector. When doing this it is an advantage to use the smallest possible entrance and exit slits. For the measurement itself the slit width must be set according to the element to be determined and the spectrum of the sample as a whole. Frequently the control of the entrance and exit slits is coupled together. If there is a possibility of coincidence with the line selected for measurement, a smaller slit width should be chosen if the sensitivity of the instrument permits. In the case of molecular radiation the slit width is less critical to the result of the analysis, and it may therefore be rather wider. This permits a lower sensitivity of the electronic equipment to be employed, with the advantage of better stability. The optimum sensitivity setting of the instrument depends on the requirements of the analysis. With instruments which are not based on the compensation principle it suffices to make sure that the intensity obtained with the maximum standard can be recorded. Before this parameter is adjusted, however, the galvanometer deflection for the pure solvent or zero standard must always be determined. In this case the measurement procedure is the same as the one described earlier.

Work with instruments based on the compensation measurement principle is rather more complicated. This group includes, for example, the Uvispek with flame photometric adaptor made by the Hilger Co. This adaptor includes a potentiometer from which a current can be branched off to compensate the photoelectric current of the detector and a sensitive galvanometer used as zero instrument. The range of concentrations to be determined is set by first compensating the dark current of the detector (no radiation falling onto the detector) by means of the respective potentiometer, then the maximum concentration standard is dispersed. The shutter in front of the detector is opened and the photoelectric current thus obtained is compensated with a special potentiometer marked "check". Changing from one potentiometer to another is achieved by means of a simple switch. The compensation of the current is indicated in all cases by the zero position of the galvanometer. The standard solution of maximum concentration is used to set 100 % transmittance and the current obtained with closed detector shutter corresponds to zero transmittance. Then the values corresponding to the individual standards and samples may be measured. For this purpose, the switch must be placed in the measurement position. For further compensation of the photoelectric current a potentiometer is used which is connected to a scale calibrated in transmittance values. During operation the two basic settings, i.e. zero and 100 % transmittance, must be checked periodically. The value corresponding to the zero standard is also noted, as it forms the beginning of the calibration curve when plotted graphically. The entire procedure of work with compensation-type instruments may be summarised as follows:

1. the instrument is switched on and allowed to stabilise
2. the respective wavelength is adjusted
3. the burner is ignited and the maximum concentration standard is dispersed
4. the wavelength setting is adjusted

4a. the slit width is adjusted

5. the zero position of the instrument is adjusted with closed shutter
6. the instrument is compensated for the maximum concentration standard and the zero setting is checked: this procedure is repeated until both values remain constant
7. the measurement is carried out and the result is evaluated.

Some instruments permit the radiation intensity to be measured by means of recording facilities. In this case the instrument automatically records a selected spectral range. This procedure has the advantage that any unexpected coincidence can be recognised from the record, and moreover the spectral background may be read directly from the record, an advantage not offered by instruments fitted with filters only. An example of an analysis with record is shown in Fig. 12.14. The background is read by connecting the background values before and after the line measured, and then measuring the height of the respective line. Obviously this procedure is more complicated than the preceding methods, as the equipment employed is more sophisticated. It is difficult to transmit general principles of operation from one type of equipment to another, and therefore the reader is referred to catalogues or instructions supplied with the instruments. However, the principle that the setting of the main operational parameters of the instrument must be checked constantly and that the measurement of standards as well as samples must be repeated, applies to every case.

10.4 Maintenance of flame photometers

The same principles apply to the maintenance of all types of flame photometers: they differ only in details. Firstly, the instrument must be kept clean. Any contamination of the suction nozzle or burner will cause the instrument to be unstable in operation. Therefore, distilled water must be dispersed after every series of measurements has been completed, until the galvanometer indicates the steady deflection value corresponding to pure distilled water. With instruments fitted with a mist chamber it is recommended occasionally to disperse dilute hydrochloric acid. When water is being dispersed the flame should be burning. When the flame is extinguished, dispersion should be stopped and air allowed to flow through until the tubing to the burner is completely dry. This decreases the extent of corrosion of the metal parts of the burner and dispersion device. If the instrument is in operation frequently it is recommended to occasionally dismantle it and wash the glass parts in chromosulphuric acid or permanganate mixture.

The burner nozzles, on which salts may crystallise if concentrated solutions are used, must be cleaned with a rough brush or washed with water.

The nozzle of the dispersion device is cleaned with a bristle or fine wire, which usually is supplied with the instrument. It is often possible to save the work of dismantling the dispersion device by using a water suction pump to flush the apparatus with air in the reverse direction to that used in normal operation. When the liquid nozzle is being cleaned the suction opening of the pump is connected to the suction capillary of the dispersion device.

In the case of the Zeiss Model III flame photometer dust is removed from the optical parts of the instrument with a soft clean piece of cotton and a soft brush, shaken free of dust. The concave mirror and the outer surfaces of the condenser are easily accessible for cleaning: the coloured filters must first be screwed out of the holder. If the outer condenser surface is dusty, the four screws on the housing (marked in red) are freed and the flanges removed. When the shutter and iris diaphragm are opened the outer condenser surface is cleaned with a hair brush without touching the closing mechanism of the shutter.

In the case of spectrophotometers, maintenance of the burner and dispersion device is the same as with the filter-type Zeiss instrument. Maintenance of the electrical part is rather more complicated, as photocells or photomultipliers are usually used: therefore the part of the apparatus containing the detector may not be opened as long as the instrument is switched on.

The electrical part of spectrophotometers is specially sensitive to corrosion, and the function of all regulators and control facilities must be checked at frequent intervals. Likewise all potentiometers must be cleaned regularly. It is advantageous to have spare parts available, mainly electron tubes, which in case of a defect cannot be replaced by local products. Repairs of the electronic part of the apparatus must, in case of a defect, be done by a specialist. Spectrophotometers also are far more exacting as to the stability of the electrical current, which again often is not very good in industrial regions. Special current stabilisers are therefore sometimes supplied together with the spectrophotometer. Perfect earthing of the instrument also is important to the stability of measurements.

When not in use, all instruments should be protected from dust by means of housings or covers of PVC foil. These are sometimes supplied by the manufacturer with the apparatus.

10.5 Preparation of solutions for analysis

10.5.1 Selection of the concentration range

In order to achieve accurate and precise work with the flame photometer, a suitable range concentrations of the element to be determined in the solution analysed must be selected. This depends on the material in which the element is being determined. As mentioned already in section 5.3, calibration curves may be deformed by self-absorption of radiation. Hence, it follows that the greater the concentration of the element analysed, the greater the curvature of the calibration curve. Fig. 10.4 shows the variation of the shape of the calibration curve according to the concentration range which is being studied.

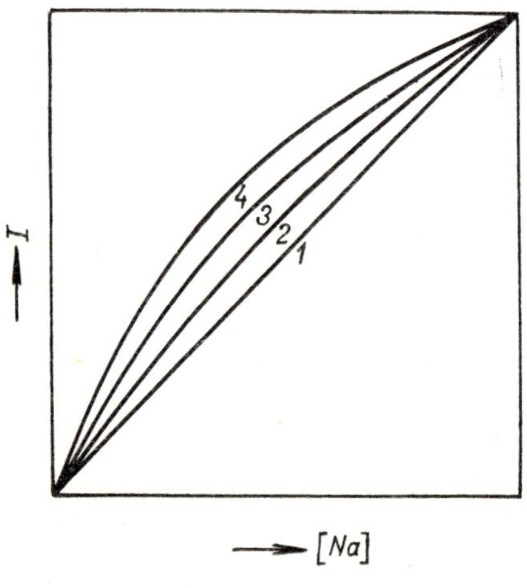

Fig. 10.3. Shape of calibration curves for the determination of sodium for various concentration ranges

1 — 0-1 ppm, 2 — 0-10 ppm, 3 — 0-100 ppm, 4 — 0-1000 ppm

Of course it is best to work with a linear calibration graph, but for practical purposes this would require too narrow a concentration range, so that we have to be satisfied with a compromise. For example, when determining alkali metals standards are usually prepared in concentrations of 0 to 10 mg in 100 ml of the solution. For some low-sensitive elements this range must be increased up to 100 mg in 100 ml. With these elements, however, the calibration curve is deformed only with substantially larger concentrations.

The concentration of the sample which is dissolved for analysis also is very important in determining the concentration range. The concentration of the solution usually should not be greater than 1 g sample in 100 ml. Greater salt concentrations would cause difficulties when the instrument has been in operation for some time. The solvent is evaporated in part from the aerosol droplets at the moment when the aerosol is being transported into the flame with the stream of gases leaving the burner. If the salt concentration in the solution is too great, the burner nozzle is clogged, the size of holes in the burner cap is reduced and the resistance of the apparatus to the gas stream is increased. This actually alters all the basic parameters of the instrument. The second unfavourable influence of the salt concentration arises from the fact that the physical properties of the solution may be altered to some extent, causing the rate of suction of the sample solution in the capillary and the rate of dispersion to alter, not to speak of other parameters as for example the degree of coagulation of the aerosol, etc. It is therefore recommended that solutions of sample and standards should have roughly equal physical properties, and therefore also roughly equal concentrations of dissolved salts. Dispersing devices with narrow capillaries, mainly including those which are directly connected to the burner, are specially sensitive to disadvantageous factors of this kind. Another unfavourable consequence of a large concentration of salts in the solution may be clogging of some parts of the apparatus, e.g. mist chamber, drop separator, etc. If the use of solutions of high salt concentration cannot be avoided, the apparatus must be dismantled after every analysis and its glass parts cleaned. For these reasons it is preferable to use a lower concentration range, though not to an extent where it is necessary to increase the sensitivity setting of the instrument above the limit of reliable operation.

10.5.2 Preparation of standards

The necessary attention must be given to the preparation of standard solutions, as the result of the analysis depends to a substantial extent on the standard solution. Pure reagents must be used in this preparation, although ultra-high purity is not essential in all cases. Considering that in most cases the mean precision of the flame-photometric analysis is not better than 2 %, it suffices to use a reagent of 99.5 % purity as basic substance for preparing the standard solution. Obviously these conditions vary from case to case: where the conditions permit better precision to be achieved, purer basic substances must be employed. Generally reagent-grade substances are used in laboratories to prepare standard solutions. If these contain water of crystallisation or if hygroscopic substances, etc., are involved, and if there is no guarantee that some changes have not taken place during storage, either other substances must be used to prepare the standards, or the content of the active component must be determined by means of some suitable analytical method, e.g. complexometric titration. With some very rare elements, where the respective reagents are not available in the required purity and their separation and determination would be too difficult, the purity of the reagent must first be checked for example by spectroscopic means and, if necessary, the reagent must be purified by crystallisation, precipitation, etc.

As far as possible, easily soluble substances are used to prepare standard solutions, while in other cases, if such substances are not easily available or they are of uncertain composition, the element may be used in the elementary form (e.g. in the case of metals): another possibility is the use of salts easily soluble in acids, e.g. carbonates, which may be dissolved by means of a suitable decomposition procedure. During the preparation process care must be taken to avoid loss or contamination of the sample.

In storage of standard solutions, attention must be given to the material of vessels employed for this purpose. For example, standard solutions of alkali metals, should not be stored in glass vessels, because alkali metals are extracted from the glass walls, increasing the content of the element analysed in the standard solution. The reverse effect is observed with, for example, dilute magnesium solutions, where the magnesium concentration is decreased by adsorption on the walls of glass vessels. In general it may be recommended to store standard solutions in vessels made of plastics, e.g. polyethylene or teflon, where extraction or adsorption practically do not occur.

It is advantageous to prepare a more concentrated stock solution of the basic substance and to dilute this as needed to prepare the standard solutions. Thus variations in the concentration of dilute solutions are avoided and there is no need to weigh the basic substance each time when a series of standard solutions is to be prepared. Sometimes it is necessary to prepare standard solutions of the same composition as that of the sample solutions. This fact must be considered when selecting the standard substance. For example when sodium is to be determined in a solution containing sodium sulphate (e.g. when silicates have been decomposed with hydrofluoric and sulphuric acid) sodium sulphate should be chosen as the standard substance: to determine sodium in the form of its chloride, sodium chloride is a suitable standard substance. If the basic substances are not pure enough and have to be purified before use, suitable procedures will be found in the literature which deals with preparation of the standard substances. In other respects the composition of the standard solution depends on the analytical procedure and method of evaluation employed.

10.5.3 Decomposition and preparation of samples

When substances insoluble in water are to be analysed practically, the sample must first be decomposed. To do this, the procedures conventionally used in chemical analysis are employed, obviously with due consideration of the suitability of the procedure selected in terms of later work with the flame photometer. From the point of view of flame-photometric analysis, the decomposition method should satisfy the following conditions.

1. It should bring the component analysed into solution completely (this condition applies to other analytical techniques also).

2. The solution obtained should not contain an excessive amount of salts, as these might interfere in the analysis proper (crystallisation in the dispersion device, burner, etc.).

3. The solution must be clear: otherwise the solid phase must be separated by filtration or sedimentation, and only the clear solution may be used for analysis.

4. The solution must not be corrosive to such an extent that it attacks any part of the apparatus.

5. Reagents and substances which might interfere by their own radiat-

ion, or otherwise influence the radiation of the element to be determined, may not be used for the decomposition process.

6. As far as possible, substances which may liberate noxious gases and vapours in the flame should not be used (e.g. fusion with potassium cyanide).

On preparation of the sample, organic substances are usually mineralized, e.g. by the KJELDAHL method, decomposition with concentrated nitric acid in the CARIUS tube, dry oxidation by means of combustion and dissolution of the ash, etc. Many different methods may be chosen to decompose inorganic substances. Silicates may be decomposed by heating with sulphuric and hydrofluoric acid, ammonium hydrogen fluoride and oxalic acid. Sulphide ores are decomposed best with aqua regia or nitric acid. Metals are decomposed with the use of various acids or their mixtures. Various special methods of decomposition have been developed for different materials. Sometimes, however, the solution obtained by decomposition cannot be used for analysis immediately. If the determination of small amounts of the analysed component is involved, it may be essential to concentrate this component preliminarily, e.g. by means of precipitation, ion exchange, extraction, zonal fusion, electroanalysis, or other techniques. In other cases again, components which may have a great influence on the determination must first be separated by precipitation, ion exchange, or other procedures. In all cases the final goal is to obtain a solution which, after concentration or separation, should be ready for flame-photometric analysis, satisfying all the above-mentioned conditions. In some cases it is advantageous to prepare solutions of standards and samples in organic solvents, which frequently increase the sensitivity of the determination. This is specially useful in those cases where the determination is preceded by concentrating the elements to be determined into an organic phase. Details of some of the above techniques will be described in the chapter on analytical procedures.

10.6 Determination of the analytical results

As in many other analytical methods, the results of flame photometric analyses are usually obtained by comparing the sample with standards, i.e. solutions containing known amounts of the element to be determined.

For this purpose a calibration curve is constructed, illustrating the relationship between the intensity of the emitted radiation at the wavelength specified and the concentration of the element analysed. The radiation intensity may obviously be replaced by some other quantity which is directly proportional to it, e.g. the galvanometer deflection (in general, the signal of the measuring equipment). The results must be noted down with great care, in order to make it quite clear which measurement belongs to which calibration curve.

When analyses are only done occasionally, normal laboratory note-books suffice for documentation of the numerical values measured and of the results. The following data must be recorded: notation of the sample, concentration and composition of the standard solutions, parameters of the apparatus, e.g. gas pressure, diaphragm value, filter or wavelength, galvanometer deflection of the zero solution, etc.; in addition the individual measured values or their means, if the measurement is repeated. The calibration curve is constructed separately on graph paper. In the case of routine analyses it is better to use preprinted forms where the individual data are written into the respective columns. A simple form of this kind is shown in Fig. 10.4.

10.6.1 Construction of the calibration curve

Let us illustrate the construction of the calibration curve and its evaluation by means of the analysis of a solution containing a single element. As an example, let us consider the determination of the NaCl concentration in an aqueous solution which contains no other component. We know that this solution may contain the element to be determined in the concentration range of 1 to 5 mg in 100 ml. We therefore prepare standard solutions containing sodium in the form of its chloride in the following concentrations: 0 mg Na, 2 mg Na, 4 mg Na, 6 mg Na in 100 ml. The instrument is put into operation and the respective filter put into its place. The galvanometer is set so as to show zero deflection for the radiation emitted by the flame itself. Then the deflection caused by a solution containing the solvent alone (i.e. water in this case), without the element to be analysed is determined.

This value is called the zero standard deflection: it is important to record it in the note-book. Now the standard solutions are dispersed one by one, starting with the lowest sodium concentration, in our case 2 mg in 100 ml, and again the measured galvanometer deflections are recorded. Now the analysis of the sample solution may be carried out.

Substance: Samples of NPK-fertiliser Date: 12. 3. 1964
To be determined: K
Analyst: A. Anton

Filter	Galvanometer sensitivity	Maximum deflection	Acetylene – air flame	
			acetylene pressure	air pressure
K 77 J	10×	450	40 mm H_2O	0.4 at

Sample notation	Sample weight	Galvanometer deviation in scale divisions	Amount read off the calibration graph mg/100 ml	Content found in %
1	3.0	250	5.67	9.5
2	5.0	303	6.80	6.8
3	2.5	358	8.00	16.0
4	4.0	160	3.72	4.6
5	3.5	210	4.80	6.9

Fig. 10.4. Example of a laboratory record

Again, the galvanometer deflection is recorded. For greater reliability of the results the entire procedure is repeated. A calibration graph is now constructed from the recorded values. On the abscissa the concentration of the element analysed is plotted on a suitable scale, galvanometer deflections being plotted on the ordinate. Plotting the respective galvanometer deflections for all the standards measured in terms of the concentrations of the element analysed, points are obtained which may be connected to form the calibration curve. This is more or less linear (Fig. 10.5). In some cases it is curved towards the concentration axis due to the influence of self-absorption of radiation in the flame, in other cases it may be curved in the opposite direction, which is usually due to ionisation of the element analysed in the flame. The curves may also be S-shaped. Therefore the concentration range for which the analysis is being carried out must be selected suitably: in the case of an S-shaped curve the solution analysed must be adjusted in an appropriate manner, as will be described in chapter 12.

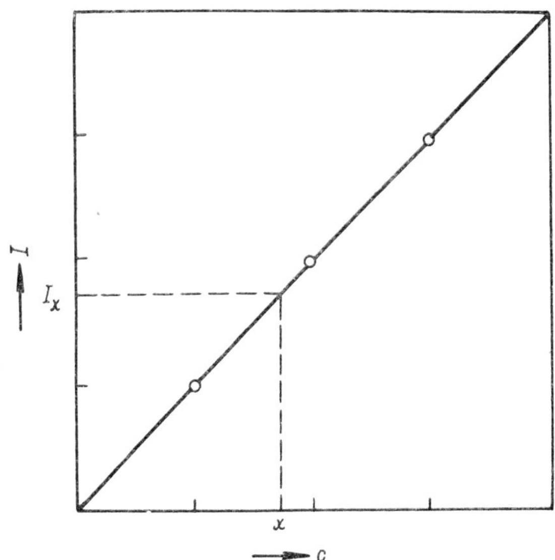

Fig. 10.5. Construction of a calibration curve and determination of the result
c — concentration, x — concentration of the element analysed,
I — intensity, I_x — intensity corresponding to the analysed solution

The concentration of the analysed element is read from the calibration curve as follows: from a point on the ordinate axis, corresponding to the measured galvanometer deflection, a line parallel to the abscissa is constructed. From the intersection of this line with the calibration curve a line perpendicular to the abscissa axis is drawn. Its intersection with the abscissa indicates the concentration of the element analysed. This procedure of measurement, evaluation, and graph construction is used for other methods also without change, although here it has been demonstrated with a solution containing a single element only. Such analytical procedures, in which the galvanometer deflection (signal) increases with the growing concentration of the element analysed, are called direct ones.

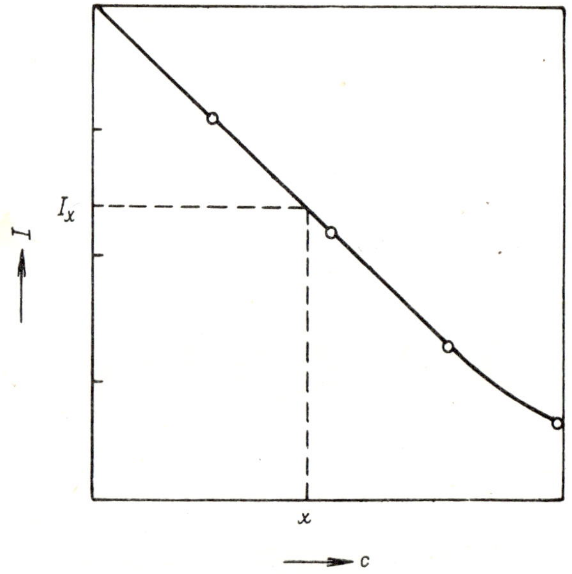

Fig. 10.6. Calibration curve for the indirect method
c — concentration, x — concentration of the element analysed,
I — intensity, I_x *— intensity corresponding to the solution analysed*

The construction of the calibration graph for an indirect method of analysis is somewhat different. In this case the standard solutions contain, besides increasing concentrations of the component to be determined, also a constant concentration of the element, the radiation of which is measured. For better illustration let us construct this calibration graph for the indirect determination of phosphates. The standard solutions are prepared so as to contain equal amounts of calcium and increasing amounts of phosphate. On measurement, a solution containing no phosphate is dispersed first, the respective deflection being measured and recorded. This is repeated with all the other standard solutions and finally with the sample. A calibration graph is constructed from the measured values corresponding to the individual standard solutions. Since in this case the intensity of calcium emission decreases with the increasing phosphate concentration, the curve has a reversed slope, as seen in Fig. 10.6.

10.6.2 Methods of sample measurement

As mentioned already, the resulting signal, measured with a flame photometer, depends on a number of different factors which cannot always be controlled exactly. Variations of oxidant and fuel flow-rates, changes in the dispersion device and mist chamber as well as changes in the electrical detection system may cause slow variations of the measured deflections. These variations may be of two kinds. Either the slope of the calibration curve changes, i.e. overall sensitivity, or the calibration curve is moved. The first case may be caused by fatigue of the detector, changes in the degree of amplification, or efficiency of the dispersion device. Movement of the calibration curve is usually due to a shift of the "zero" of the instrument. When measuring standards and samples we obviously try to eliminate these unfavourable influences as far as possible.

The simplest, as well as most frequently used procedure is to check the zero position of the instrument and one of the standard solutions after every second, third, or fifth sample, adjusting all conditions in a manner suitable to re-establish values corresponding to the calibration curve. The number of samples after which the deflection corresponding to the standard and zero solutions is checked, depends of course on the stability of the instrument and the required accuracy of measurement.

There is, of course, a second possible procedure, i.e. calculating the ratio of the sample deflection to the deflection obtained with one of the standards, instead of adjusting the sensitivity of the instrument. The calibration curve is in this case plotted in the same ratio, and the procedure is called the quotient method. The standard is usually measured after two samples, and its value for any given sample is interpolated from two consecutive measurements. This procedure is more laborious but also more accurate than the preceding one, although it is incapable of eliminating errors caused by a shift of the zero position of the instrument.

With the third, most laborious procedure, the samples are first measured in an approximate way and then, immediately one after the other, the sample and two standards of roughly the next higher and lower concentration are measured, the sample being measured between the two standards. In this case the result need not be calculated graphically, as the following equation can be used

$$c_x = c_1 + \frac{I_x - I_1}{I_2 - I_1}(c_2 - c_1), \tag{10.1}$$

where I_1, I_2 and I_x are the deflections corresponding to the standards and the analysed sample, c_1, c_2 and c_x are the respective concentrations. Obviously the concentrations of the standards must be sufficiently close to each other as to allow the calibration curve to be replaced by a straight line in this region. With this procedure, errors due to a shift of the zero position as well as errors caused by variations of the sensitivity of the instrument are eliminated. This procedure is sometimes referred to as the bracketing method [413].

10.7 Procedures of evaluation for eliminating interfering effects

When the technique of flame photometry is used for analytical purposes, it happens only rarely that an element is determined in a simple solution: generally, solutions of practical samples also contain other components, the presence of which may have a greater or lesser influence on the radiation of the element analysed, depending on the character of these components. If the concentration of the required element were determined by means of a calibration curve which had been constructed with the use of standard solutions containing none of the interfering components, substantial errors would ensue. These problems are discussed in detail in chapter 12. Some of the interfering influences may, however, be eliminated by preparing suitable standard solutions or by choosing a suitable procedure of evaluation. Many such methods have been suggested: here we shall mention only those which are used most often in practical analysis. The selection of the method is mainly influenced by the properties and composition of the sample itself. Let us discuss the method of model standards first.

10.7.1 The method of model standards

With this method, errors are eliminated by using standard solutions which contain the same interfering components as the solutions analysed. In both cases the interfering components will influence the analysis in the same way, so that the result of the analysis remains unaffected. This method is specially suitable for routine analyses of technical material, the composition of which is relatively constant and known. This condition is usually satisfied by the majority of raw materials, semi-products and final products occurring in practice. If the composition is not known beforehand, it can be determined by one of the conventional analytical methods and the standard solutions are then prepared according to the result of this analysis. Obviously all components of the analysed solution need not necessarily cause interference. Therefore it suffices to include in the standard solution those components which are known to influence the analysis in some way. SCHUHKNECHT recommends an even simpler procedure. He suggests that the concentration of the interfering components be determined in an orientative manner only by flame photometric measurement, using the respective filters, and then to prepare the standard solutions according to the values thus obtained.

10.7.2 The method of calibration curves

This procedure is the method of model solutions extended for routine analysis of samples of rather varying composition containing, however, a relatively constant amount of the element to be determined. SCHUHKNECHT [687] developed a procedure which avoids this difficulty to a certain extent, although the basic principle of the method remains. The procedure is based on the use of a system of calibration curves. To analyse a certain type of material, several series of standard solutions, each containing a different concentration of the interfering components are prepared. As an example, let us take the determination of sodium and

Table 10.1. STANDARD SOLUTIONS

Calibration series	Notation of standard solution	Composition of standard solution	
		mg K_2O/1	mg Na_2O/1
1	1 K 1	50	0
	1 K 2	40	0
	1 K 3	30	0
	1 K 4	20	0
	1 K 5	10	0
	1 K 6	0	0
2	2 K 1	50	50
	2 K 2	40	50
	2 K 3	30	50
	2 K 4	20	50
	2 K 5	10	50
	2 K 6	0	50
3	3 K 1	50	100
	3 K 2	40	100
	3 K 3	30	100
	3 K 4	20	100
	3 K 5	10	100
	3 K 6	0	100
4	4 K 1	50	200
	4 K 2	40	200
	4 K 3	30	200
	4 K 4	20	200
	4 K 5	10	200
	4 K 6	0	200
5	5 K 1	50	500
	5 K 2	40	500
	5 K 3	30	500
	5 K 4	20	500
	5 K 5	10	500
	5 K 6	0	500

potassium. The composition of standard solutions is seen in Table 10.1. The concentration of the interfering element is first determined with the use of the flame photometer in a preliminary manner, and that series of standard solutions is then used in which the concentration of the interfering element is closest to the content of this element in the sample. The position of the calibration curves corresponding to the individual series of standards differs to some degree, as seen in Fig. 10.7, where the calibration curves for the determination of potassium in the presence of sodium are shown. These standard solutions may be prepared beforehand in sufficient amounts to serve in the analysis of the different sample types. Recently, SCHUHKNECHT [689] widened this principle even further by suggesting a system of calibration solutions which should permit the determination of alkali metals in any material whatever.

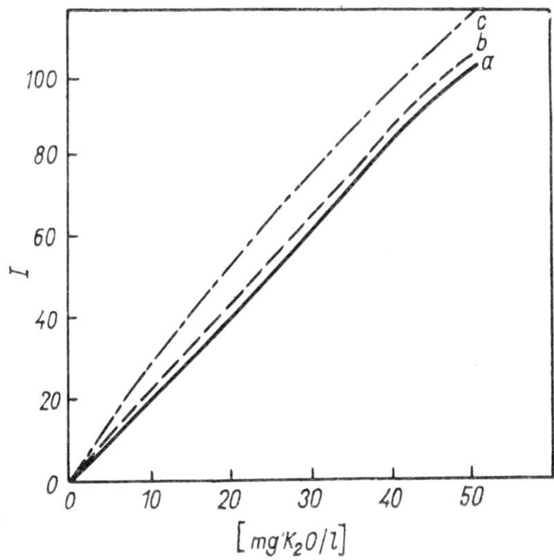

Fig. 10.7. System of calibration curves
I — intensity, a — K_2O + 0 mg Na_2O/l, b — K_2O + 50 mg Na_2O/l, c — K_2O + 500 mg Na_2O/l

10.7.3 The method of correction curves

If the concentration of the interfering element varies only little, the entire course of the calibration curve with different concentrations of the interfering element need not be determined: it suffices to determine accurately the content of the interfering element and to make the respective correction in the individual measurements. The correction value is read from a correction curve, constructed by plotting the error of the analysis against the concentration of the interfering element, usually in units of the absolute amount of the element to be determined. If radiant interfering influences only are involved (see section 12.4) a single correction curve suffices for different concentrations of the element analysed. If more complicated interfering influences are involved, several correction curves usually have to be plotted for different contents of the element to be determined. This procedure was employed, for example, to correct the influence of potassium on the determination of rubidium [397].

10.7.4 The standard addition method

When samples of greatly variable composition are to be analysed, or samples of which only a small number are to be analysed, the standard addition method will be found more useful [48], [67], [114], [196], [263], [281], [283], [284], [326], [339], [388], [556], [649], [650], as it does not require any laborious preparation of model standards. This method is suitable for the determination of elements in samples which differ substantially from one another and in the analysis of which the formation of volatile compounds, viscosity of the solution due to the presence of organic substances, etc., interfere. The method is based on the assumption that the analysed element which is present in the original sample is influenced in the same way as another known amount of the same element which is intentionally added to the sample, and that furthermore the calibration curve passes through the origin of the coordinate system. The radiant interference caused by imperfect isolation of the measured line from the rest of the flame radiation cannot be eliminated in this way.

When this method is applied, a certain shape of the calibration curve is assumed. In the simplest and most conventional case, it is assumed that the calibration curve is linear and that therefore the only unknown constant which has to be determined is the slope of the calibration curve. To determine this, a single addition suffices. The ratio of the amount of the element added to the amount originally present is important for the precision of the analysis. In the ideal case the two amounts are roughly equal. In this case the deflection caused by the original concentration is one half the deflection after the addition. The required concentration is calculated from the relation

$$X = \frac{I_x \cdot a}{I_{a+x} - I_x}, \tag{10.2}$$

where X is the required amount, a the amount added, I_x, I_{a+x} are the measured deflections before and after the addition.

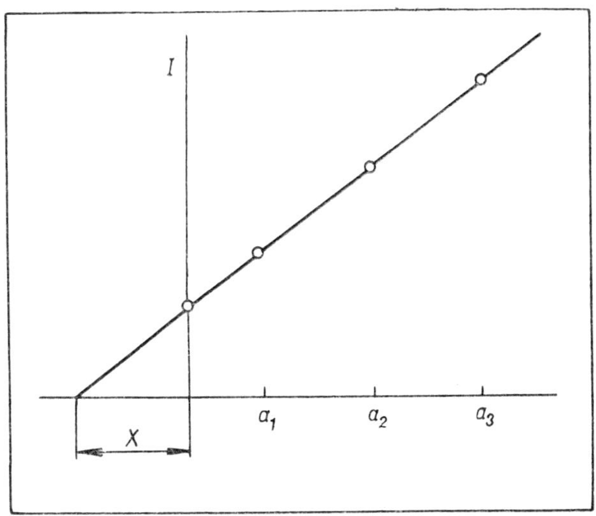

Fig. 10.8. Evaluation by means of the standard addition method

a_1, a_2, a_2 — *additions in mg, X — content of the element to be determined in the volume measured, mg*

To verify the linearity of the calibration graph and to improve the precision of the analysis, several additions may be used. Evaluation is done by graphic means in this case, as shown in Fig. 10.8. The overall volume of the solutions must be kept constant, i.e. water must be added to the sample to which no standard solution is added. As example, let us describe the determination of strontium in water according to CHOW and THOMPSON [136], [137].

The standard solutions are prepared in the same way as with other procedures. Five times 50 ml of the water sample are measured by pipette

50 ml distilled water being added to the first aliquot and 50 ml of standard solutions of increasing strontium concentrations to the other aliquots. The measured values are plotted on a graph, from which the strontium concentration is read as shown in Fig. 10.9.

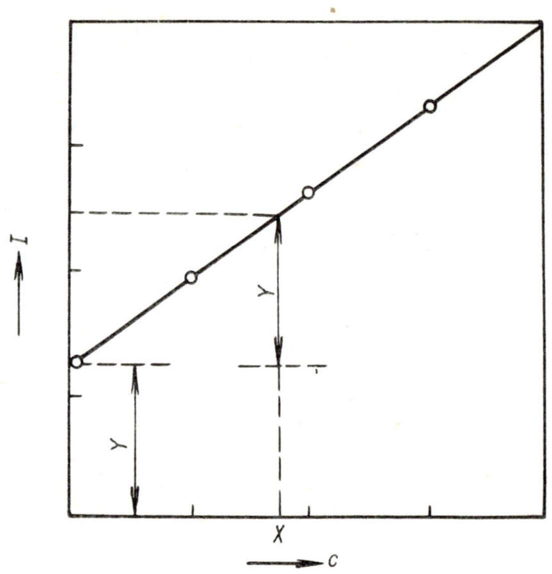

Fig. 10.9. Evaluation by the Chow and Thompson procedure
c — concentration, I — intensity, Y — intensity of the solution analysed, X — required concentration

It is always essential to subtract the background from the individual deflections carefully, as otherwise the resulting concentration of the element analysed would be found to be greater than the true value. For this reason also, this method of evaluation is advantageous in atomic absorption, where there is less danger of radiant interferences and the calibration graphs are usually linear over a limited region.

If the calibration graphs are curved, a larger number of additions must be used to calculate at least two constants which characterise the form o fthe curve. Evaluation may then be done by numerical or graphic means, with the use of tabulated values or nomograms. Although the process of evaluation itself is not very complicated, its theoretical justification would take up too much space: therefore, the reader is referred to the original literature [67], [607].

10.7.5 The internal standard method

Another method of eliminating some interfering influences is the internal standard method [44], [62], [63], [64], [89], [128], [135], [172], [173], [263], [364], [445], [538], [597], [640], [710]. It was originally worked out for spectral analysis, where it is used nearly exclusively, but it has also been applied to flame photometry. The internal standard is an element which either is not present in the original sample at all, and equal amounts of this element are added to the sample and standards, or it may be one of the elements contained in the sample, the concentration of which in individual samples varies very little. This internal standard must satisfy a number of other demands too. Firstly, its behaviour in the flame must be identical with the behaviour of the element to be determined. It must have roughly the same ionisation potential. It must emit radiation, the excitation energy of which does not greatly differ from the excitation energy of the radiation studied, and this radiation must not coincide with the measured radiation of the element analysed. As examples, let us use the case of lithium serving as internal standard for the determination of sodium, caesium in the determination of rubidium, etc. Strontium is unsuitable as a standard for the determination of sodium if phosphoric acid or aluminium is present in the sample, etc. When all the above demands are satisfied, we may assume that the ratio of intensities of the two elements, or of corresponding values, will be independent of any possible influence, and result of the analysis will be free of error. This procedure is advantageous in those cases where changes of the physicochemical properties of the solutions are involved, such as are difficult to determine. However, radiant interference influences usually cannot be eliminated in this manner. It is not essential for the calibration graph to be linear in this case. The internal standard must be chosen very carefully, considering the influences which are to be eliminated. Otherwise, new errors might be introduced into the analysis.

For better illustration, let us describe the determination of sodium with the use of lithium as internal standard. The solutions of standards and samples are prepared in the conventional way. Each of the sample solutions, however, also contains the same amount of lithium: e.g. for a series of standards containing 0, 2, 4, 6 mg Na, 5 mg Li may be added. The galvanometer deflections are determined for the sodium radiation, and then the galvanometer deflections of the lithium radiation are recorded for every standard, obviously using the respective interference filter or with the correct wavelength set on the monochromator. The ratio of these values is then calculated, i.e. $V_1/V_2 = Q$, V_1 being the galvanometer deflection corresponding to sodium and V_2 the deflection corresponding to lithium. The values of Q are plotted on the ordinate axis: in constructing the calibration curve, concentration values are usually plotted on the abscissa. It should be noted that the compound of the internal standard added should not itself contain the element to be determined.

10.8 The technique of work in absorption flame photometry

Although the working techniques in atomic absorption are very similar to those of the emission method, there are some peculiarities which we shall mention here. As in emission flame photometry, standard solutions are used to construct calibration curves, from which the required concentration of the sample solution is then read. For the preparation of standard solutions, the same applies as in the emission method. Calibration curves are obtained by plotting absorbance (optical density) against the concentration of the respective element. The following procedure is used for the measurement itself. The discharge lamp is first switched on and checked so that its radiation falls onto the detector. The flame and dispersion device are adjusted. With the diaphragm of the instrument closed, i.e. when no radiation from the lamp falls upon the detector, the zero position of the instrument is set and a specified instrument deflection (usually the maximum) is adjusted with open diaphragm. If the instrument deflection is set with the use of the amplifier or voltage applied to the dynodes of the photoelectric multiplier, the zero setting of the instrument must usually be re-set, as now the dark current (or the noise) of the multiplier has been altered. The two processes, i.e. setting of the zero value and of the full deviation, are repeated until the required values are obtained in both cases. Since there is usually no danger of radiation interference in atomic absorption, it is advantageous to set the full deviation by means of the diaphragm or monochromator slit. In this way, laborious corrections can be avoided. When the instrument has been set in this way, individual standards are measured in the order of increasing concentrations and the respective absorbance values are calculated according to the formula

$$A = \log \frac{I_0}{I} \tag{10.3}$$

where I_0 and I are the deflections corresponding to the pure solvent and the respective standard.

Even with very good stability of the instrument employed, this will allow only concentrations of two orders of magnitude to be measured, i.e. transmittances of 10 to 99%. While the concentration range measured may be varied by the use of different degrees of amplification rather easily in emission flame photometry, this possibility does not exist in absorption photometry. Therefore, the path of radiation through the flame is usually shortened to measure high concentrations, i.e. the burner is rotated to let the radiation pass through the flame cross-wise instead of longitudinally (this corresponds to the use of shorter cells in spectrophotometry), or another line of the element analysed, with a lower oscillator strength, is used. Thus for example, sodium may be determined by measuring the doublet Na 3303 Å, the oscillator strength of which is roughly 70 times less than that of the yellow doublet: therefore, the sensitivity obtained is 70 times lower. To increase the sensitivity of the measurement on the other hand, either the path of the radiation in the flame must be prolonged by using multiple passing (see the section on instruments) or the respective absorbance must be "dilated" over the entire scale by means of compensation equipment. In this case the amplification is increased and the absorbance of the most concentrated standard is compensated down to the zero value of the galvanometer scale by means of a voltage in opposite direction to the current of the photoelectric multiplier. In this case the zero value of the instrument lies outside the scale, and there is no possibility of determining the absolute values of I_0 and I needed for the calculation of the respective absorbance value. In this case the calibration graph may be plotted from the relation

$$c = K(I_0 - I) \tag{10.4}$$

where c is the concentration of the element in the solution, I_0 the deflection obtained with the pure solvent, I are deflections obtained when the respective standards are dispersed. This relation is not perfectly linear, but it is close to it.

11. Accuracy and sensitivity of the method

11.1 Accuracy

When the determination of the same element in the same sample is repeated several times, the results obtained will not necessarily be equal: the individual measured values may differ from each other to some extent. The variation of these values determines the precision of the method. Generally it may be assumed with full justification that with a determination repeated many times over the results would exhibit a normal (GAUSSIAN) distribution around the mean value. This distribution is characterised by a parameter σ, which determines the "width" of the

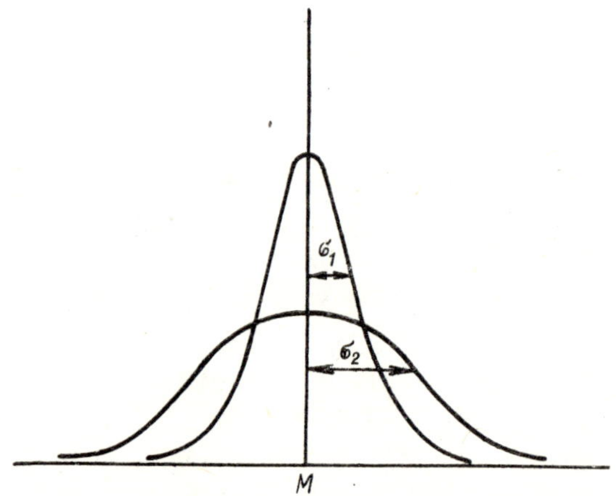

Fig. 11.1. Normal distribution of the results of measurement
M — true value, σ_1 — standard deviation of the more precise method, σ_2 — standard deviation of the less precise method

Gaussian curve. A low standard deviation, σ, indicates a low degree of variation, i.e. high precision, while a large value of the standard deviation indicates that the precision of the method is low and its variation is high. The two possibilities are illustrated in Fig. 11.1.

The accuracy of a method is as important as its precision: accuracy means agreement of an experimental result with the true content of the element analysed. The standard deviation is a measure of the precision of measurement, but it does not allow us to find out whether the measured result is accurate or whether it is subject to a systematic error. A systematic error may be elucidated by analysing a reference standard in which the content of the element analysed is known with great reliability, for example by comparing the results of flame-photometric analysis with those of a different technique. Sometimes, inter-laboratory control analyses may be used for this purpose. A more detailed discussion of this problem would be outside the scope of this book, and the reader is referred to the specialised literature [226], [834]. In the following we shall assume that our results are accurate but imprecise.

The value of the parameter σ may be calculated from repeated determinations. Since usually rather a small number of determinations are used to calculate this value, the result is only an estimate which depends on a random selection of data. Obviously the greater the number of determinations, the closer is the result to the true value of σ. Since only an approximate value is involved, we denote it S to differentiate it from the true value, σ. The standard deviation may be calculated from repeated measurements according to the formula

$$S = \sqrt{\frac{\sum_i (x_i - \bar{x})^2}{n - 1}} = \sqrt{\frac{\sum x_i^2 - n\bar{x}^2}{n - 1}} \tag{11.1}$$

where x_i it is the value of the individual measurements, $\bar{x}$ the mean value obtained from the relationship $\bar{x} = \frac{\sum x_i}{n}$, n is the number of measurements. When this formula is employed, the value of n should not be less than 10. The expression $n - 1$ is called the number of degrees of freedom.

The standard deviations may also be calculated from analyses carried out only twice in parallel, if a larger number of samples of similar composition is being analysed. This procedure is sometimes more convenient from a practical point of view. In this case the following formula is applied

$$S = \sqrt{\frac{\sum_i (x_i - x_i')^2}{M}}, \tag{11.2}$$

where x_i and x_i' are the two results of analyses of a single sample, M is the overall number of determinations. The number of degrees of freedom is $M/2$ in this case.

Since, in this case, the concentration of the element analysed in the individual samples may differ, and the values of x_i differ accordingly, the difference between the two determinations must be converted to relative values of $\frac{x_i - x_i'}{(x_i + x_i')/2}$ before substituting into the relation. The standard deviation thus calculated is already expressed in relative values.

The relative standard deviation expressed in percentages is sometimes called the variation coefficient, which is denoted by the letter V or C. It is defined by the relation

$$C = \frac{S}{\bar{x}} \cdot 100 \tag{11.3}$$

In flame photometry it is advantageous to express the relative standard deviation in percentages, since this is practically independent of the concentration of the element analysed, except of course in extreme cases of very great concentrations or concentrations close to the sensitivity limit.

Especially with a small number of determinations, the standard deviation may be estimated by an even simpler method, using the variation range, i.e. the difference between the largest and smallest result of parallel determinations. This is calculated from the relation

$$S = k_n \cdot \bar{R} \tag{11.4}$$

$\bar{R}$ being the variation range, i.e. the difference between the largest and smallest result, k_n a constant which depends on the number of parallel determinations: this constant is tabulated in Table 11.1. This method of calculation is only used if the number of analyses is small, generally less than 10.

Table 11.1. VALUES OF k_n FOR CALCULATING ESTIMATES OF STANDARD DEVIATIONS FROM VARIANCE RANGE VALUES

n	k_n	n	k_n
2	0.886	7	0.370
3	0.591	8	0.351
4	0.486	9	0.337
5	0.430	10	0.325
6	0.395		

The estimate of the standard deviation may also be employed to calculate the reliability interval, i.e. the interval in which the error of the determination will lie with a selected probability value: from this, the concentration range may be obtained in which the true result will lie with a selected probability. This calculation is mainly of significance in those cases where an element is being determined which should be contained in the sample with a concentration greater or lesser than a certain specified value, i.e. in a type of analysis which is usually carried out in technical control laboratories of industrial plants. It is possible to find out in this way whether a raw material or a product is of standard quality, or whether its quality is doubtful. If a large probability value is chosen, a wider interval is obtained, and vice versa. In analytical practice, a probability (generally called statistical security) of 95 % is selected, sometimes it is 99 % or exceptionally even 99.7 %. Table 11.2 lists the coefficients by which the standard deviation must be multiplied to give the reliability interval corresponding to the selected statistical security. Since the standard deviation calculated is an estimate only, being the more accurate the greater the number of measurements from which the value was calculated, this coefficient also depends on the number of measurements.

If a determination is repeated several times and the mean value of the results $\bar{x}$ is taken, the standard deviation of the arithmetic mean is defined by the relation

$$S_{\bar{x}} = \frac{S}{\sqrt{n}} \tag{11.5}$$

Table 11.2. VALUES OF THE COEFFICIENT t

Number of degrees of freedom	Statistical security		
	95 %	99 %	99.7 %
1	12.71	63.66	235
2	4.30	9.92	19.2
3	3.18	5.84	9.22
4	2.78	4.60	6.62
5	2.57	4.03	5.51
6	2.45	3.71	4.90
7	2.37	3.50	4.53
8	2.31	3.36	4.27
9	2.26	3.25	4.09
10	2.23	3.17	3.89
15	2.13	2.95	3.54
20	2.09	2.85	3.38
25	2.06	2.79	3.29
∞	1.96	2.58	2.97

where $S_{\bar{x}}$ is the standard deviation of the mean value, S the standard deviation of a single determination, n the number of measurements from which the mean was calculated.

As an example, let us describe the calculation of the standard deviation and statistical security of the determination of rubidium by the flame-photometric technique (Table 11.3).

The standard deviation is calculated from twenty analyses each twice repeated, including the chemical decomposition of the sample.

The reliability interval for a statistical security of 95 % is obtained by multiplying the standard deviation by the respective coefficient from Table 11.2 for a number of degrees of freedom of twenty.

The resulting interval is

$$(X \pm X \cdot 5.9 \cdot 2.09/100) = (X \pm X \cdot 0.123\,\%).$$

11.2 Sensitivity

In flame photometry, as with other methods, sensitivity denotes the lowest concentration or amount of the element which can be determined. Con-

Table 11.3. CALCULATION OF THE STANDARD DEVIATION

Rb content found, %		Difference, %	
1. determination x_1	2. determination x_2	$\frac{x_1 - x_2}{\bar{x}} \cdot 100$	$\left[\frac{x_1 - x_2}{\bar{x}} \cdot 100\right]^2$
0.19	0.18	5.4	29.16
0.15	0.14	6.2	38.44
0.19	0.19	0	0
0.080	0.072	11.2	125.44
0.25	0.27	7.7	59.29
0.16	0.17	6.05	36.03
0.11	0.10	9.5	90.25
0.15	0.15	0	0
0.16	0.16	0	0
0.18	0.19	5.4	29.16
0.14	0.14	0	0
0.080	0.082	2.5	6.25
0.091	0.086	5.7	32.49
0.064	0.082	24.5	600.25
0.073	0.081	10.3	106.09
0.11	0.10	9.5	90.25
0.073	0.082	11.6	134.56
0.083	0.080	3.2	13.69
0.082	0.081	1.2	1.44
0.084	0.082	2.4	5.76
$S = \sqrt{1398.55/40} = 5.9\,\%$			Σ 1398.55

centration data are generally expressed in terms of the solution dispersed. This conventional definition of sensitivity depends not only on the condition of the measuring apparatus, but also on a number of other factors, so that it differs with one element from one case to another. For the sake of precision, the declared sensitivity is always given together with the respective experimental conditions, although it nonetheless also includes the author's subjective view. Obviously this definition is rather unsatisfactory, although it has become conventional and is often used. The concept thus defined, however, has no precise meaning. Therefore, analysts have in recent times stopped expressing the least concentration which can be determined in the above manner, replacing it by the accurately defined term "limit of detection": this has recently come to be used in emission flame photometry, while it is already being widely used in atomic absorption. The reason is that the minimum amount of the element which still can be detected, can readily be determined by diluting the solution analysed only as long as the element in question can be determined. To determine this value accurately, it is recommended to start from the standard deviation of the blank experiment [426].

In flame photometry the blank experiment value, i.e. the value obtained when dispersing a blank sample containing none of the element to be analysed, is determined by the dark current of the photoelectric multiplier, by the light dispersed in the optical system, and by the flame background or interference of molecular bands or lines of other components of the solution. The variation of this value is expressed by the standard deviation σ_{bl}.

When determining very low concentrations, the value of the blank experiment is substracted from the measured value corresponding to the sample (or this value may be compensated directly by electrical means). With respect to the additivity of squares of standard deviations, the standard deviation of the blank experiment influences the resulting accuracy of the determination in both cases. Since small values, close to the blank experiment value, will also have an equal standard deviation, we may write

$$\sigma_x = \sqrt{\sigma_{\text{an}}^2 + \sigma_{\text{bl}}^2} = \sigma_{\text{bl}}\sqrt{2},$$

where σ_{bl} is the standard deviation of the blank experiment, σ_{an} is the standard deviation of the signal corresponding to the element to be determined, σ_x is the overall standard deviation of the analysis.

To decide whether an element is present or not, a high statistical security value is usually chosen — 99.7 %, corresponding in Table 11.2 to a coefficient of 3 (for a standard deviation accurately known). Thus the value measured must be greater than $3\sqrt{2}\,\sigma_{\text{bl}}$ in order that we may be able to state with a probability of 99.7, that the element in question is present in the sample. The concentration corresponding to this value is at the same time the limit of detection of the method employed.

$$x - x_{\text{bl}} = 3\sqrt{2}\,\sigma_{\text{bl}} \tag{11.7}$$

where x is the signal corresponding to the minimum concentration which can be determined, x_{bl} is the signal corresponding to the blank experiment.

The concentration which corresponds to this instrument reading is, at the same time, the limit of detection of the given element. The value of the limit of detection may therefore be decreased by decreasing the standard deviation of the blank experiment, or by increasing the deviation of the measuring instrument which corresponds to the respective element concentration. Although, especially in atomic absorption, these values are being used increasingly, it must be kept in mind that actually this characteristic was up to now used in qualitative analysis. In fact it states how small an amount of a specific element can be detected (determined qualitatively), not determined (quantitatively). In the preceding discussion we did not consider the error of the result, so that a concentration equal to the limit of detection would, for example, be determined with an error of $\pm 100\,\%$, which obviously is a totally unacceptable result in most cases. Nonetheless, the limit of detection is very important, since for example it permits inter-comparison of different lines or procedures, and under some conditions it also allows an estimate of the concentration which can readily be determined. Let us call the latter the limit of determination, defined as the lowest concentration which can be determined with a required accuracy. (Obviously this accuracy should not be greater than the accuracy of the method itself.) This value is also influenced by the properties of the instrument, sample, etc., but its statistical character is maintained. When it is stated, for example, that iron may be determined in a specified material under specified experimental conditions down to a concentration of 1 p.p.m. (the limit of determination) with an accuracy of $\pm 5\,\%$, we have a statement which sufficiently characterises the method described.

There remains now to define the sensitivity of the method. According to SPECKER [717a] this may be defined by the expression dx/dc, x being the measured value and c the concentration of the element to be determined. In other words, the sensitivity of the method states the manner in which the measured value rises with concentration, i.e. it is the slope if a calibration curve is involved. This manner of expressing sensitivity has already become conventional in atomic absorption, where sensitivity is defined as a concentration change which causes an absorption of 1 %, corresponding to an absorbance of 0.0044. Similarly, DEAN [177] defined the sensitivity of emission flame photometry as the concentration change causing a transmission of 1 %.

Besides these methods, which may be assumed to be generally valid, other authors have defined sensitivity of the limit of detection in different ways. For example, GILBERT [373] defines, with respect to the BECKMAN spectrophotometer, the limit of detection of a specified element at a given wavelength as a concentration which causes a deviation of the measuring instrument equal to 1 % of the deviation of this instrument corresponding to the flame background, at the optimum setting of the optical system and flame. BUELL [104] defines sensitivity as double the value of the background (up to a background value which is a maximum of 20 % of the entire scale).

12. Interfering factors causing erroneous results

Flame photometry is no absolute method, it rather compares emission (absorption) values of the sample with values determined for standards. With respect to the complicated nature of processes which take place in the analysis, there is a danger that with small differences in the composition of the sample and standards the results will be incorrect. All factors which may influence the determination in this manner are called interfering influences, and much attention has always been given to them.

In order to clarify the ways in which interfering influences may originate, let us start with a review of the processes which take place in flame photometry (see Fig. 12.1). Each individual step is influenced by a number of factors, some of which may vary in an uncontrolled manner in the course of an analysis. The factors which vary in this way may influence the resulting concentration of that form of the element to be determined, which is capable of emitting or absorbing radiation. In the absorption method these are free atoms, in the emission method atoms, molecules, or radicals which are capable of being excited. Let us now discuss the individual processes and mention the factors which may have an interfering influence.

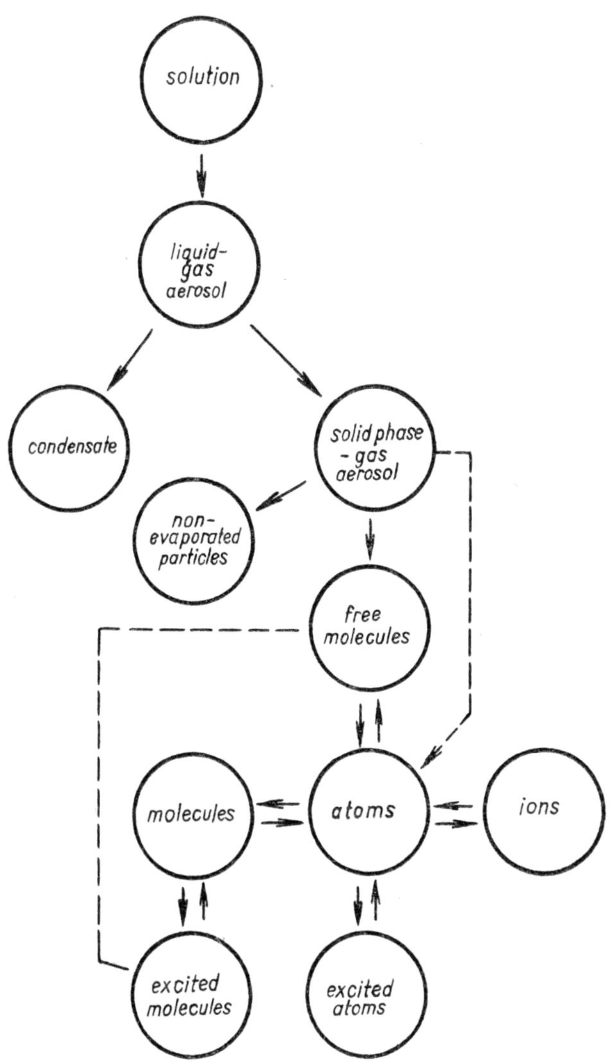

Fig. 12.1. Diagram of processes taking place in flame photometry

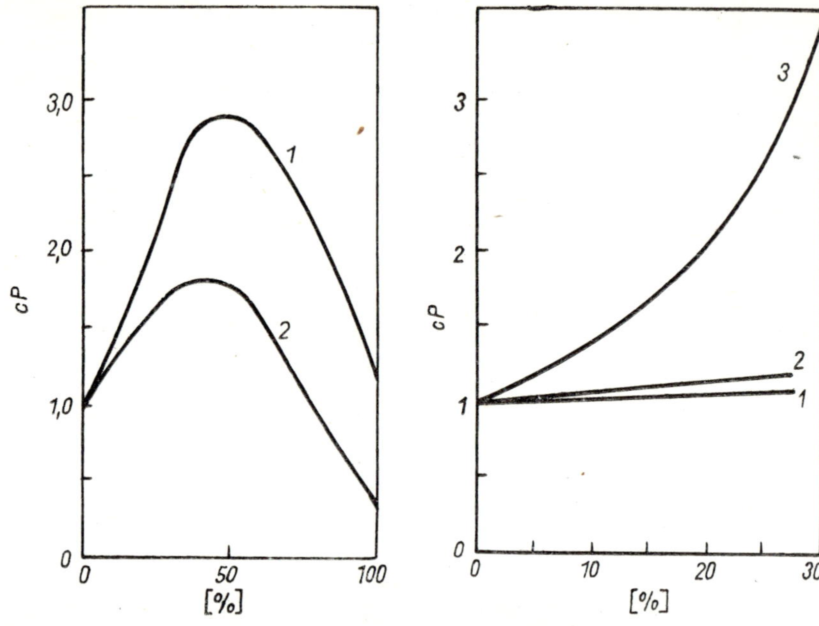

Fig. 12.2. Viscosity of some solutions of organic substances in water
1 — ethyl alcohol, 2 — acetone

Fig. 12.3. Viscosity of aqueous solutions of NaCl (1), LiCl (2), sugar (3)

12.1 **Interfering influences of transport and their elimination**

This group first of all includes the interfering influences which act upon the dispersion process, and also coagulation and sedimentation in the mist chamber. In dispersion the most important factor is the amount of the solution analysed which is dispersed in a unit of time, and the resulting size of droplets, which later influence the processes of coagulation and sedimentation. As was already shown in section 7.2, viscosity mainly determines the amount of solution dispersed with constant parameters of the dispersion device. Among substances which may influence the viscosity of a solution, sugars, urea, and some other organic compounds should be considered when biological materials are being analysed. With organic solvents mixed with water, viscosity as a rule depends on the composition of the solution (see Fig. 12.2): the viscosity maximum, with a given composition of the solution analysed, will be manifested by a minimum rate of dispersion (Fig. 7.1). Inorganic salts likewise increase the viscosity of the solution, althought to a much smaller degree. Fig. 12.3 shows the relationship between the viscosity of aqueous sugar and sodium chloride solutions and their composition. The amount of solution dispersed may be determined by experimental means: some authors recommend that corresponding corrections be applied.

The influence of surface tension of solutions is somewhat less unambiguous. A decrease of the surface tension causes the droplet size to decrease, and sedimentation and coagulation to slow down, so that the amount of solution which passes through the flame increases. With direct-injection burners the droplet size is determined by the time interval needed for complete evaporation of the droplet, or the number of droplets which pass through the flame without evaporating completely. Obviously these processes influence the concentration of free atoms in the flame.

Inorganic salts increase the surface tension to some degree, in proportion to their concentration. The effect is the greater the greater the charge of the ion involved and the smaller its radius. Therefore, care must be taken when analysing solutions with relatively large concentrations of salts, and the possibility should be investigated of using standards with lower salt concentration. On the other hand, organic substances generally decrease the surface tension. Fig. 12.4 shows the relationship between surface tension and composition, for some frequently employed solvents.

All factors which influence the amount of solution entering the flame in a unit of time, may be eliminated nearly completely by using the internal standard method. With the majority of commercial instruments, however, there are difficulties because the instruments are not equipped for the simultaneous measurement of two lines. An exception is the PERKIN – ELMER Model 146 flame photometer, which is fitted with apparatus for simultaneous measurement of a lithium line. Lithium was chosen because it is a relatively rare element and is not present in the majority of materials analysed by flame photometric means. With respect to other influences which may interfere at the same time, it should be mentioned that lithium is suitable as internal standard for a small number of elements only, and that other elements may also be employed as internal standards. In every case the suitability of this method for the given case must be considered. The internal standard method is used regularly in flame spectrography, since it eliminates the influence of the photographic process at the same time. In this case we are free to choose the element to serve as internal standard and the respective line practically with no limitation.

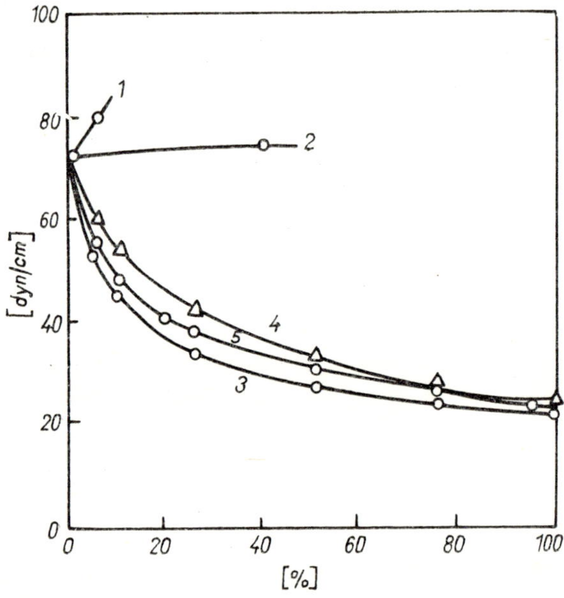

Fig. 12.4. Surface tension of some aqueous solutions of organic substances and of NaCl
1 — NaCl, 2 — sugar, 3 — ethyl alcohol, 4 — acetone, 5 — methyl alcohol

To eliminate the interfering influences which act upon the solution before it enters the flame, another procedure of general validity may be selected, for example the method of standard additions (10.7.4).

In so far as none of the methods mentioned is used, it is essential to keep all the properties which influence transport processes on a constant level. These are mainly viscosity, surface tension and vapour pressure of the solvent. This means practically, that it is essential to keep constant concentration of solvents added, for example, to increase the sensitivity of the determination, as well as the content of other organic substances, e.g. sugars, urea, etc., in the case of analysis of biological material. When inorganic materials are being analysed, the main substances to be considered are chelate-forming reagents and, to some extent, the overall content of inorganic salts.

12.2 Interfering influences of the solid phase and their elimination

From the moment when the solvent evaporates from aerosol droplets and the particles are formed in the solid phase, coagulation stops being an important factor. The evaporation of solid particles, however, takes place at relatively high flame temperatures. In this process—conversion from the solid to the gaseous phase—several types of interfering influences may participate. Most often it is the so-called quenching effect: this is caused by the fact that some components of the solution analysed decrease the radiation intensity of the element studied, or they decrease its absorption value. The reverse phenomenon occurs very rarely, i.e. the presence of only few components of the solution analysed causes the emission intensity or absorbance of the element determined to increase. From the practical point of view only the first effect, i.e. the quenching effect, is important. The second effect occurs sporadically to such a degree, that its influence is very limited and we are mentioning it only in order to make the description of interfering influences complete from the theoretical point of view.

The quenching effect is caused by incomplete evaporation of aerosol particles in the flame, the consequence of which is a lower concentration of free atoms in the flame, and therefore a lower signal results. The magnitude of this effect depends on the size of solid aerosol particles which are formed when the solvent has evaporated. The quenching effect is more distinct in the case of instruments with direct aerosol inflow, where the mean size of particles entering the flame is greater. When a mist chamber is used larger particles are separated by sedimentation and they coagulate. Fig. 12.5 illustrates the decrease of the emission intensity of calcium radiation, caused by an addition of colloidal silicic acid [261] with the two types of apparatus. The degree to which particles evaporate in the flame may be illustrated by comparing the emission intensity of the element analysed in the basic solution and in the same solution diluted in a ratio of 1 : 1. When particles evaporate completely, emission decreases to 50 %, while with incomplete evaporation it decreases relatively less.

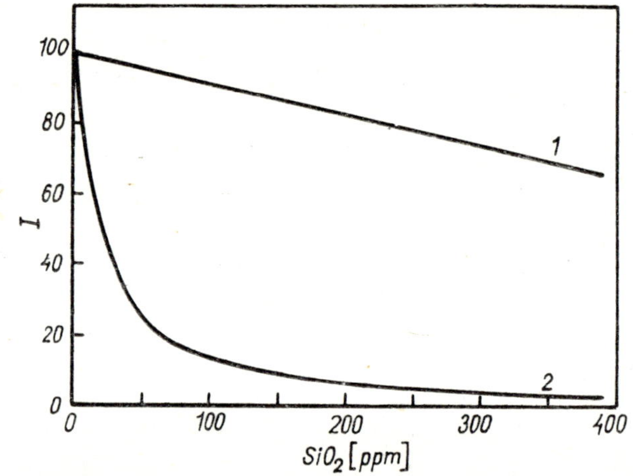

Fig. 12.5. Influence of colloidal SiO_2 on the emission intensity of calcium
I — intensity, 1 — apparatus with mist chamber, 2 — apparatus with direct aerosol injection into the flame

When the solution is diluted, the resulting aerosol particles are smaller after evaporation of the solvent, and evaporation is more complete. In the extreme case the flame intensity decreases to only 63 % of the initial value since, as follows from relation 7.1, the relationship between intensity and concentration is defined by the expression

$$I \sim \frac{\mathrm{d}m}{\mathrm{d}t} \sim r^2 \sim c^{2/3} \tag{12.1}$$

The validity of this relation is based on the assumption that no other influences depending on the concentration of the element analysed participate in the process, e.g. ionisation interferences or self-absorption. This relationship is also corroborated by some experimental data. For example, LEYTON [475] has found that in the case of calcium phosphate the calibration curve is bent, with an exponent of 4/5, while with calcium chloride it takes a regular course. RUBEŠKA, et al. [655], found in the analysis of sodium in limestone by means of atomic absorption that the slope of calibration curves decreases with the calcium concentration to the 2/3 power.

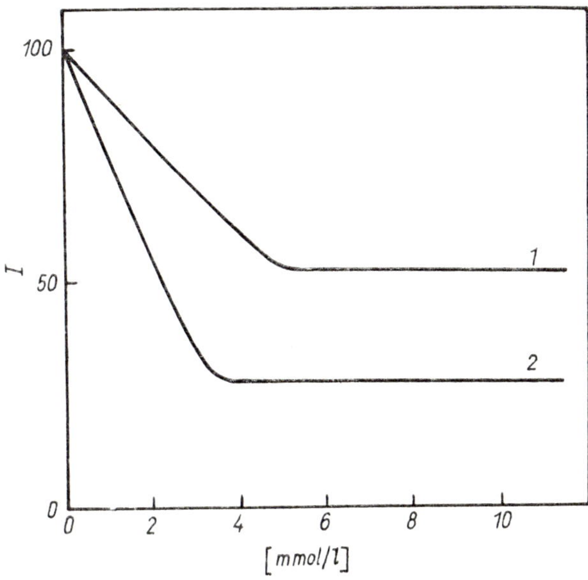

Fig. 12.6. Influence of sulphate and phosphate anions on the emission intensity of calcium
1 — sulphate, 2 — phosphate, I — intensity
Calcium concentration — 5 mmol

The interfering influence mentioned is mainly significant in the determination of alkaline earth elements in the presence of a number of compounds, including sulphates, phosphates, vanadates, molybdates, tungstates, arsenates, oxalates, borates, and furthermore aluminium, iron, chromium, zirconium, titanium, thorium, uranium, beryllium, and others. When the relationship between the emission intensity of the element analysed and the concentration of the interfering component is plotted, a curve of a characteristic shape is obtained. In the presence of some components emission decreases down to a certain value, and does not vary any more with further addition of the interfering component. The form of a relationship of this type is illustrated in Fig. 12.6. Frequently, the molar ratio (element analysed/interfering element) is plotted on graphs instead of the concentration of the interfering component. This type of relationship was found when studying the emission intensity of alkaline earth elements, for example in the presence of phosphates, sulphates, vanadates, arsenates, and chromates.

In many other cases rather a different shape of the relationship between emission intensity of the element analysed and the concentration of the interfering component was found. In the graphic representation, no breaking point was observed in the curves, and the emission intensity was found to decrease constantly with the rising concentration of the interfering component, as shown in Fig. 12.7. This type of relationship was found in the presence of aluminium nitrate, beryllium, silicic acid, titanium and zirconium.

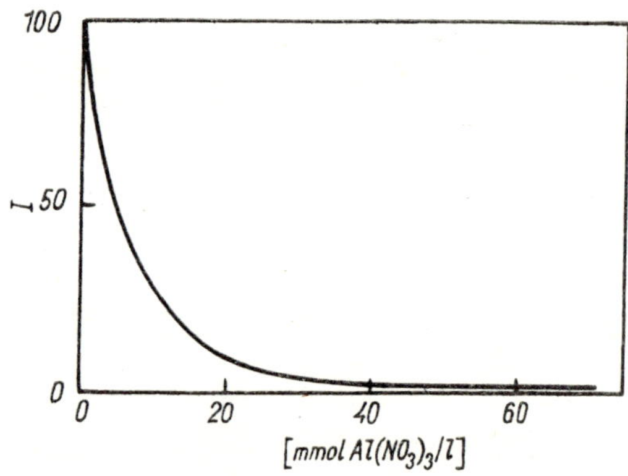

Fig. 12.7. Influence of aluminium on the emission intensity of calcium
Ca concentration — 5 mmol, I — intensity

This difference of behaviour may be explained mainly by the differing boiling points of the oxides of the interfering elements, as shown by ALKEMADE and VOORHUIS [13], [14], and SHUHKNECHT and SCHINKEL [686].

The first type of interference is explained by the fact that the element analysed forms a compound with the interfering component, the boiling point of this compound being high to such a degree that thermal decomposition of the compound in the flame is not complete, and thus the concentration of free atoms in the flame is decreased [38]. In aerosol particles, the excess oxide of the interfering element (which itself is volatile) evaporates quickly after the solvent has evaporated, and in the second phase only that amount of the interfering element remains which is bound in a compound with the alkaline earth element to be determined, this compound being thermally more stable. Of the original conglomerate, only those compounds remain which have a high boiling point or point of decomposition. Depending on the flame temperature, however, a part of this thermally stable compound evaporates also, liberating a part of the element analysed: this is manifested by "residual emission", which corresponds to the maximum intensity decrease. Depending on the thermal stability of the compound and the flame temperature, residual emission may be greater or lesser. The molar ratio in which the element analysed forms a compound with the interfering element may be determined by the method of continuous variations, as used in spectrophotometry. For example, the following ratios were found: Ca[Sr] : Zr – 1, Ca : P – 1, Ca/Cr – 1, Ca : U – 3/2 [590]. Thus the amount of the interfering component which suffices to decrease the emission intensity of the element analysed is less than stoichiometric, while on the other hand an amount greater than stoichiometric has no more effect under conventional conditions.

The interfering components which we have included in the second group are distinguished by a higher boiling point of their oxides, so that evaporation of the excess oxide from the particles takes place at a slow rate. The interfering component, which forms the matrix of the sample, is less volatile than the aerosol particles which were formed from the pure solution of the substance analysed. In this case the decrease is mainly caused by occlusion (blocking) of the component analysed in particles formed from the interfering component which, due to its relatively high boiling point and thus also large particle size, does not evaporate in the flame. The rising amount of the interfering component thus also increases this blocking and decreases the emission intensity of alkaline earth elements with no limitation. As examples of this type of interference, let us mention observations by PUNGOR and KONKOLY–THEGE [615] with respect to the determination of Sr in excess Ca, and the observations of RUBEŠKA, et al. [655], on the determination of Na in limestone.

It follows from the above description of the mechanism of this process, that the second type of interference will also take place when a thermally stable compound of the two components, analysed and interfering, is formed. This effect also takes place when calcium is determined in the presence of aluminium nitrate. Aluminium oxide does form defined compounds with a relatively high point of evaporation with calcium oxide, but in the presence of excess aluminium oxide formed from the nitrate the effect due to blocking participates to a larger extent. The same applies to the other above-mentioned components also. In every case there must be an excess of the interfering component. In some cases, the difference between the two groups may be quantitative only, depending on the flame temperature to a large degree. For example, vanadium, which we have included in the first group, belongs to the second group in terms of its behaviour in flames of lower temperature [213].

It was proved in a number of studies that slightly-volatile compounds are formed when solid aerosol particles evaporate, and no process taking place in the gaseous phase is involved. When a system of two dispersion devices is used to disperse separately the solution containing the interfering component, the quenching effect will not occur.

For the sake of completeness, let us also mention the interference which occurs on conversion from the solid to the gaseous phase, where, however, the emission intensity or absorption value of the element analysed is increased. In this case a compound is formed which is more volatile than the compound formed by the element analysed in the original sample solution. As an example, let us mention the influence of fluoride ions on the emission intensity of aluminium [452]. The volatility of the fluorides of this metal is greater than that of other compounds. Some authors believe that the effect of oxine on the emission intensity of aluminium is similar in character.

Finally, a case may also occur where a volatile matrix causes an increase in the value of the signal observed, without itself forming a compound with the element analysed. This effect may be explained by the fact that the slightly-volatile matrix contributes to the formation of small aerosol particles and thus also to easier evaporation of the element analysed. ALKEMADE [17] mentions, as a possible example, the influence of salts of aliphatic acids on the emission intensity of calcium.

The quenching effect may be reversible in character, i.e. the two components, analysed and interfering, influence each other in the same manner. Thus, for example, the presence of boric acid in the solution analysed has an unfavourable influence on the emission intensity of boron. It is to be seen that this case will occur when the two components form a compound which only evaporates in the flame with difficulty.

The quenching effect, however, need not be reversible in character, particularly in those cases where the interfering component and the element analysed do not form a compound. In this case the more volatile component, which itself may be blocked by the less volatile one, may cause an increase in the volatility of the less volatile one, if it is present in excess.

Some authors have observed another effect which also belongs to the group of effects taking place on conversion from the solid to the gaseous phase. When magnesium was determined by means of atomic absorption in a propane-air flame, it was found that the presence of calcium, strontium, and to a lesser extent, sodium, causes an increase in the magnitude of the signal [223], [334]. This phenomenon is explained by the fact that magnesium oxide forms a eutectic mixture with the oxides of other metals, the melting point of this mixture being substantially lower than that of magnesium oxide alone: thus, the evaporation of magnesium is facilitated. HALLS and TOWNSHEND [334] also consider the possibility of a reaction taking place between calcium in the gaseous state and magnesium oxide, magnesium being liberated in the gaseous state. These hypotheses indicate the fact that probably all processes which take place in the flame are not yet explained in a satisfactory manner, which after all is readily understood in view of the complicated nature of these processes.

Interfering quenching effects have an unfavourable influence on the analytical result, and therefore increased attention must be given to them, in order to avoid an unduly large error in the analysis. A number of methods has been suggested for their elimination, but it cannot be said beforehand, which method will be the best in a given case. Generally it is necessary to ensure the elimination of an interfering influence by experimental means. Interfering components may be removed from the solution by, for example, the use of ion exchange resins. Obviously this necessitates an additional chemical operation and the time required for the analysis is prolonged. Ion exchangers were used to remove interfering components by, for example, DAVID [167] (removal of P in the determination of Sr and Ca) and BIECHLER [69] in the determination of trace elements in industrial water. Phosphates may be precipitated with, for example, $FeCl_3$ [257], $ZrOCl_2$ [719], or $LaCl_3$ [804], [830].

In some cases it suffices to alter the degree of oxidation of the interfering element. For example, the interfering effect of vanadates is substantially weakened by the reduction of pentavalent vanadium to the tetravalent state [633]: the same applies to the reduction of hexavalent chromium to trivalent [217]. Since, in principle, the ratio of the rate of evaporation to the rate of passage of aerosol particles through the flame is involved, it sometimes suffices to carry out the measurement in the upper part of the flame, where evaporation of the particles may be complete. This very simple and undemanding procedure is usually applied in those cases where the interfering component is not present in molar excess, or where the compound formed is relatively highly volatile. Fig. 12.8 illustrates the relationship between the emission intensity of calcium and the sulphate concentration, measured with different flame heights [285]. The interfering influence of aluminium, which forms less volatile complexes, is eliminated in this way to a very slight extent only.

To decrease the quenching effect, several organic reagents have been recommended and tested, e.g. oxine [184], [817], EDTA [185], [596], [795], salicylic acid, and acetylacetone [772]. Although in most cases the quenching effect is not eliminated completely, at least it is substantially decreased. The optimum concentration of the organic reagent and the limits of its applicability must be tested by experimental means in every case, since usually the results cannot be transferred uncritically from one type of analysis to another.

Some authors explain the suppression of the quenching effect by the formation of a complex of the alkaline earth element with the organic reagent, which is then easily decomposed in the flame [795]. Other authors assume that rather the compact character of the solid aerosol particles is disturbed in the flame due to combustion of the organic substance. The increased surface then permits more rapid evaporation [38]. This view is corroborated by the fact that when direct-injection burners are used many organic reagents are suitable for this purpose, including those which have no chelating properties at all, or which have them in a limited degree only. For example, it was found in the case of

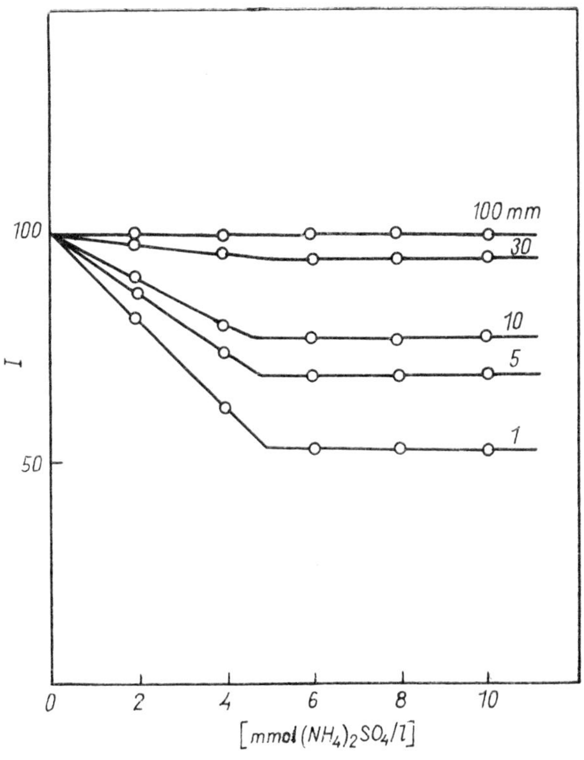

Fig. 12.8. Influence of sulphate ions on the emission intensity of calcium at different flame heights

I — intensity

the determination of magnesium that an addition of 10 % isopropyl alcohol totally suppresses the influence of phosphates, while the influence of sulphates is suppressed up to a concentration of 500 p.p.m., and that of aluminium up to 200 p.p.m. [794]. Similarly, a 2 % glycerine solution suppresses the quenching effect of phosphates on calcium [348]. Some authors also recommend a 10 % glycerine solution [624].

Another possibility of eliminating the interfering influence is liberation of the element analysed from a slightly-volatile compound by the addition of an element which forms more stable compounds with the interfering compound, or which is added in a great excess (the so-called releasing effect). For example, the interfering influence of sulphates, phosphates, and aluminium in the determination of calcium may be suppressed by additions of strontium, lanthanum, neodymium, samarium, yttrium, and iron. Magnesium, beryllium, barium, and scandium are only partly efficient [194]. The selection of one of these elements is controlled by the aspects of radiation interference, molar efficiency, and the price of the respective reagent. In some cases the excess of the releasing element must be very great. For example, it is recommended to add lanthanum in a fivefold molar excess compared to phosphate in order that the release of calcium, strontium, or barium be complete [833]. On the other hand, when calcium is used to eliminate the interfering influence of aluminium on magnesium, the required calcium excess is variable, rising with the increasing aluminium concentration [657].

The releasing effect is very frequently utilised in emission as well as in absorption flame photometry. The exact composition of the compounds formed is usually not known. Some more definite information is available for the formation of diphosphates of alkaline earth elements only [832].

This method of suppressing interfering influences may be combined with the internal standard method. For example, when calcium is to be determined in the presence of various interfering components, a relatively small excess of strontium may be added to the analysed solution, the intensity of the calcium line being compared with that of the strontium line. This eliminates the interference, since phosphates, sulphates, silicic acid, aluminium, iron, and uranium have a similar influence on both elements [263]. Obviously, either a line or a band must be measured in the case of both elements, and corrections must be made for radiative interferences. The concentrations of the two elements in the solution should be nearly the same.

12.3 Interfering influences caused by a shift of equilibrium in the flame, and their elimination

As already stated in chapter 7, a number of processes take place in the flame influencing the concentration of free atoms or of other particles in the flame. All the factors which influence any one of these processes may distort the results of the analysis. Since absorption as well as emission in the flame is proportional to the concentration of the respective particles, these influences affect both methods in the same manner.

In principle, the following processes and their states of equilibrium are involved:

1. dissociation of molecules
2. formation of new compounds with combustion gas components
3. ionisation

All these equilibria may be characterised by the equilibrium constants of the respective reactions for temperatures specified by the flame temperature. In connection with this, it should be mentioned that in as far as instruments with mist chambers are used, and the resulting flames are laminar, this temperature is well defined in the volume of the flame outside the reaction zone. In the case of instruments with direct injection of the sample solution, and with a turbulent flame, same of the particles contain the non-evaporated solvent when entering the flame. Heat and concentration gradients form around the droplets, disturbing the state of equilibrium in the flame, so that characterisation by means of some efficient equilibrium constant is highly inaccurate.

Dissociation equilibria have been investigated in greatest detail in the case of alkali metals. Therefore, we shall discuss these in the following, keeping in mind that similar relationships apply to other elements. Let us consider an arbitrary alkali halide. When molecules have volatilised in the flame, they dissociate, the process being expressed by the relation

$$MeX \rightleftarrows Me + X \qquad (12.2)$$

where Me is the alkali metal and X is the halide. The form of the dissociation process may be characterised by the use of the dissociation constant

$$K_{dis} = \frac{P_{Me} \cdot P_X}{P_{MeX}}, \quad (12.3)$$

where P is the partial pressure of the component denoted by the respective subscript, expressed in atmospheres. The value of the equilibrium constant depends on the flame temperature. Table 12.1 shows the values of the dissociation constants of alkali halides for a temperature of 2500 °K [846]. Similar relations apply to other anions also, only nitrates do not have this effect. In their case, the anion decomposes rapidly.

In the case of flames with temperatures of over 2000 °K it may be assumed that dissociation of alkali halides is complete. This means that for example in the acetylene – air, acetylene – oxygen, and oxygen – hydrogen flames, the interfering influence of these anions will not be felt. With cooler flames, e.g. town gas – air, propane – air, or butane – air, this influence of the halide must be taken into account.

Table 12.1. DISSOCIATION CONSTANTS OF SOME ALKALI HALIDES [AT.] AT $T = 2500$ °K

	Cl	Br	I
Na	$2.2 \cdot 10^{-4}$	$1.4 \cdot 10^{-3}$	$3.2 \cdot 10^{-2}$
K	$7.9 \cdot 10^{-5}$	$5.4 \cdot 10^{-4}$	$7.2 \cdot 10^{-3}$
Rb	$4.7 \cdot 10^{-5}$	$3.8 \cdot 10^{-4}$	$5.8 \cdot 10^{-3}$
Cs	$2.6 \cdot 10^{-5}$	$2.3 \cdot 10^{-4}$	$4.0 \cdot 10^{-3}$

This interference may readily be eliminated by maintaining a constant concentration of the respective anions in all samples as well as standard solutions, e.g. by adding excess acid, etc.

Similarly, other dissociation processes may also be manifested. Especially important are those processes in which a part of the compound is, at the same time, a component of the combustion gases (see

Table 12.2. PARTIAL PRESSURE OF COMBUSTION GAS COMPONENTS OF ACETYLENE—AIR FLAME [AT] [308]

N_2	0.668	H_2	0.030
H_2O	0.090	OH	0.002
CO_2	0.071	H	0.004
CO	0.136	O	0.0003
O_2	0.0003	NO	0.0008

Table 12.2). This mainly involves the formation of hydroxides, oxides, or hydrides. Any change of the equilibrium concentration of these components will obviously influence the concentration of free atoms. For example, formation of hydroxides was proved with alkali metals [409]. The tendency to form hydroxides decreases in the series Li, Cs, Rb, K, Na. In the acetylene—air flame, lithium is dissociated to 21 %, potassium to 43 % and sodium is dissociated completely [378]. Lithium exhibits a slight dependence on the pH value of the solution, as seen from

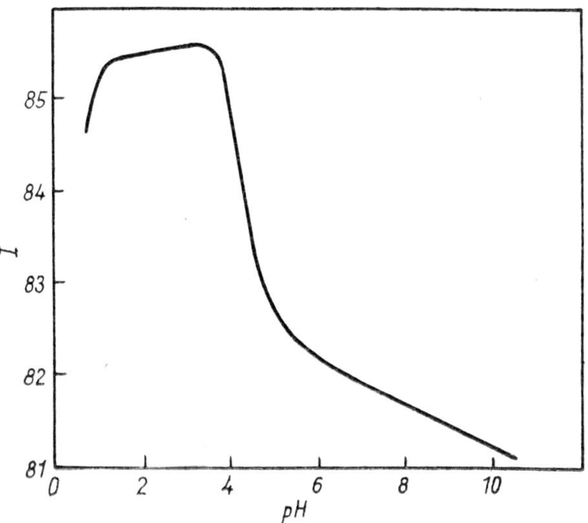

Fig. 12.9. Influence of pH on the emission intensity of lithium in an oxygen-hydrogen flame, burner with direct aerosol injection into the flame

Fig. 12.9. The influence of the pH value of the solution on the intensity of emitted radiation is more distinct in the case of the alkaline earth elements, a fact which also corresponds to their substantially lower degree of dissociation in the flame. In the acetylene – air flame, magnesium is dissociated to 1.5 %, calcium to 4.7 %, strontium to 11 % and barium only to 0.21 % [378]. Fig. 12.10 is a graph of the relation between the intensity of the strontium line 460.7 nm and the pH value of the solution according to HORR [383]. It is not yet quite clear to what extent alkaline earth metals form hydroxides as well as oxides in the flame. JAMES and SUGDEN [410] believe, from analogy with the spectra of alkaline earth halides, that the intensive molecular bands of the alkaline earths correspond to hydroxides. This is also indicated by the isotope shift observed when deuterium is used in the dispersed solution [465]. However, it appears that this discussion is not yet concluded definitely [648]. VEITZ and GOURVICH [779], [780] state that the concentration of alkaline earth hydroxides in the flame is less than 10 % of the concentration of oxides.

With the majority of elements, however, the formation of oxides is most important. Sometimes this tendency is strong to such an extent that either it makes the flame-photometric determination totally impossible or it at least substantially decreases the sensitivity of the determination. The formation of oxides is distinctly manifested in the so-called flame profile, i.e. the dependence of the free atom concentration on the distance from the burner mouth. This concentration is best estimated by means of absorption measurement, since the emission is strongly temperature dependent, so that the measured profile is distorted by the temperature distribution in the flame.

Obviously, burners with mist chamber and direct-injection burners have different profiles. In the first case, solid particles enter the flame, so that the greatest concentration is usually achieved in the reaction zone or in its vicinity. The concentration then decreases rapidly or slowly. Fig. 12.11 shows this relationship for a number of elements according to measurements by ZELUKOVA and POLUEKTOV [843]. In order to avoid loss by diffusion the author used a silica tube in the form of a chimney.

With direct-injection burners, the flame profile is determined by two processes. In the lower part of the flame the free atom concentration rises,

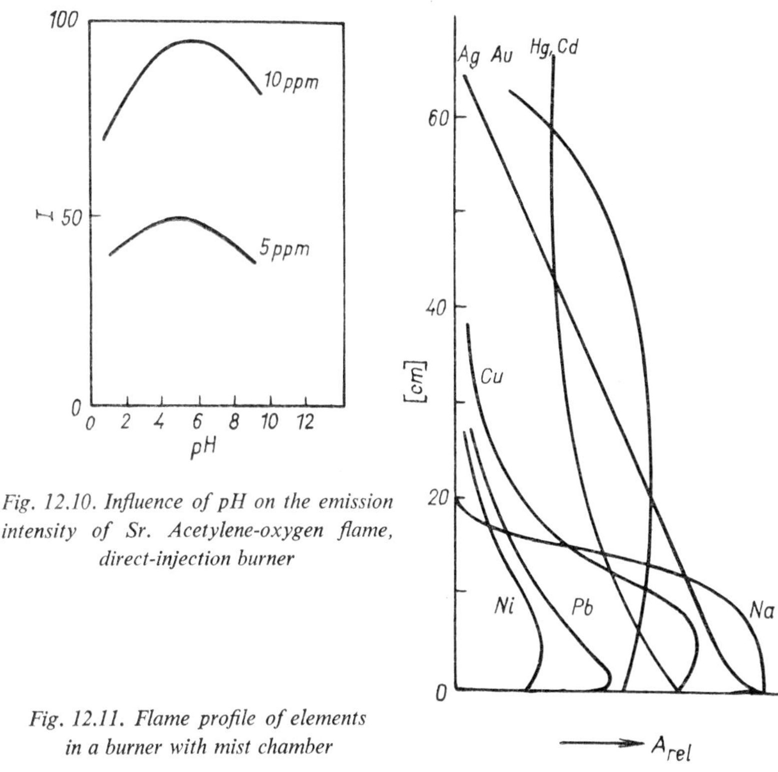

Fig. 12.10. Influence of pH on the emission intensity of Sr. Acetylene-oxygen flame, direct-injection burner

Fig. 12.11. Flame profile of elements in a burner with mist chamber

due to the decomposition of compounds to free atoms: this is followed by a decrease due to oxidation of the free atoms to oxides. The result of these two contrary processes is a concentration peak at a certain height of the flame (see Fig. 12.12). The greater the tendency of the atoms to form oxides, the lower is the peak located, and vice versa. ROBINSON [642], [645] compared the height of the peak above the burner mouth with the equilibrium constants of the reaction

$$Me + xO \rightleftarrows MeO_x,$$

estimated at a temperature of 3000 °K. Although the values of the constants were only approximate, as the values of thermodynamic functions are not known for such high temperatures, the agreement between the

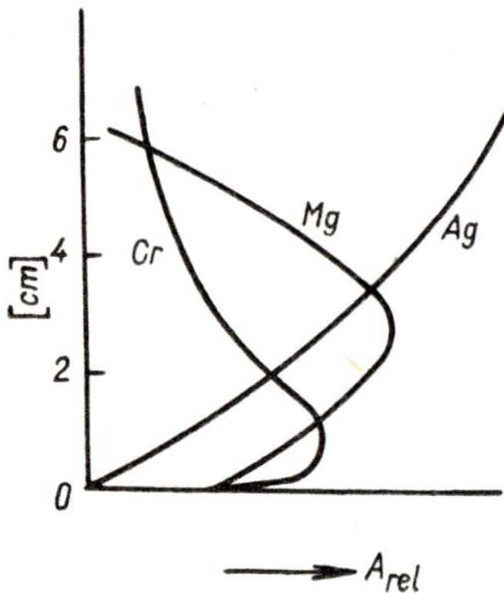

Fig. 12.12. Flame profile in the case of a direct-injection burner

calculated and measured order was perfect in the case of Cd, Ag, Tl, Mg, and Cr.

In some cases the influence of organic substances also is related to this process of formation of free atoms. If atoms of the element analysed form a complex with the organic reagent, volatilisation is easier, since the organic reagent burns easily, leaving a free atom. Thus liberation of the atom is far quicker than in the case of volatilisation from an aqueous solution, where the atom is usually present in the form of a hydrated ion, and the removal of hydration water demands the supply of a certain amount of energy and, therefore, more time. Thus, a greater concentration and therefore also a greater intensity of emitted radiation, or greater absorption, is achieved by speeding up the formation of free atoms.

When elements which have a tendency to form oxides are being determined, it is advantageous to use weakly reducing flames. For example, the sensitivity of the absorption determination of molybdenum and tin was increased by using a small fuel excess [296], in this case acetylene. It is recommended to ajust the conditions so as to make the inner flame cone a diffuse one and for white radiation to appear along the entire surface of the burner mouth. With the emission method this possibility is usually limited by the continuous spectrum which such a flame emits. Sometimes, the radiation of the reduction zone of the flame, which is a strongly reducing medium, is used directly for measurement.

DEAN and CARNES [180] studied arsenic, antimony, and bismuth lines in this way. They used light conductors to isolate radiation from the sharply limited small part of the flame in which the radiation intensity is maximum [124]. Since in this part of the flame the flame temperature is less than in the flame mantle, it must be assumed that the energy needed to excite the respective radiation is supplied by chemical reactions (chemiluminescence).

To eliminate interferences which are related to dissociation equilibria in the flame, constant conditions must be maintained in the flame: i.e. the fuel and oxidant flow-rates, as well as the amount of organic substances introduced into the flame, if any, as the latter also influences the oxidation-reduction equilibrium and spatial distribution of atoms in the flame [309].

Another process in the flame which may also influence the free atom concentration is ionisation. In general it may be expressed by the equation

$$\mathrm{Me} \rightleftarrows \mathrm{Me}^+ + \mathrm{e}^- \tag{12.4}$$

where e^- denotes a free electron, Me^+ is an ion. The ionisation equilibrium which is established in the flame may be expressed by means of the ionisation constant

$$K_{\mathrm{ion}} = \frac{P_{\mathrm{Me}^+} \cdot P_{\mathrm{e}^-}}{P_{\mathrm{Me}}}. \tag{12.5}$$

The dependence of this ionisation constant on temperature is given by the so-called SAHA equation, which takes the following form after calculation of the constants for alkali metals:

$$K_{\mathrm{ion}} = 3.23 \cdot 10^{-7} \cdot T^{5/2} \cdot 10^{-\frac{5040 E_i}{T}} \quad [\mathrm{at}], \tag{12.6}$$

where T is the absolute temperature and E_i the ionisation potential, eV. It becomes clear from SAHA's equation that the ionisation constant depends, to a large extent, on the ionisation potential of the element. The ionisation potentials and values of the ionisation constant of some elements in the town gas–air and acetylene–air flames are listed in Table 12.3. All other elements have ionisation potentials greater than 6 eV, so that their ionisation in flames is negligible. It is seen from the

Table 12.3. IONISATION POTENTIALS AND IONISATION CONSTANTS

Element	E_i eV	K_{ion} [at]	
		1970 °K	2360 °K
Li	5.36	$9.34 \cdot 10^{-13}$	$3.02 \cdot 10^{-10}$
Na	5.12	$4.16 \cdot 10^{-12}$	$1.0 \cdot 10^{-9}$
K	4.32	$4.67 \cdot 10^{-10}$	$4.9 \cdot 10^{-8}$
Rb	4.16	$1.17 \cdot 10^{-9}$	$1.1 \cdot 10^{-7}$
Cs	3.87	$6.30 \cdot 10^{-9}$	$4.47 \cdot 10^{-7}$
Ca	6.09	—	$3.55 \cdot 10^{-11}$
Sr	5.67	—	$2.81 \cdot 10^{-10}$
Ba	5.19	—	$2.8 \cdot 10^{-9}$

expression for the ionisation constant that the degree of ionisation also depends on the partial pressure of electrons in the flame. From experimental measurements it follows that the partial pressure of electrons is roughly 10^{-8} at. in a pure acetylene–air flame, and that it does not practically influence ionisation under conventional analytical conditions: [379], [484]. This source of electrons may, therefore, be neglected.

The dissociation and ionisation equilibria in flames are obviously interconnected, so that the calculation of the partial pressure of the individual components is very complicated. As an example, let us mention the equilibria which are established when an alkali halide is introduced into the flame. The following equilibria are involved:

$$K_{MX} = \frac{P_M P_X}{P_{MX}}; \qquad K_M = \frac{P_{M^+} \cdot P_{e^-}}{P_M}; \qquad K_X = \frac{P_X \cdot P_{e^-}}{P_{X^-}} \qquad (12.7)$$

where M is the respective alkali metal and X is the halide. These equilibria are moreover bound by the condition of electrical neutrality

$$P_{M^+} = P_{e^-} + P_{X^-} \qquad (12.8)$$

and the condition of a constant overall metal and halide concentration

$$P_{\Sigma M} = P_{MX} + P_M + P_{M^+} \qquad P_{\Sigma X} = P_{MX} + P_X + P_{X^-} \quad (12.9)$$

In this description, equilibria with the components of the flame have been neglected. In a simple case, when only one component is present in the solution dispersed, dissociation and ionisation depend on its final concentration in the flame. This again is determined by the initial concentration in the solution, efficiency of the dispersion device, and volume of the combustion gases. Although these parameters may vary rather widely, the relationships observed by individual authors differ by a maximum of one order of magnitude.

For example, HINNOV [377] and POLUEKTOV [587] have found the relation $P_{\Sigma M} = C \cdot 2 \cdot 10^{-4}$, JAMES and SUGDEN [409] $P_{\Sigma M} = C \cdot 2{\cdot}5 \cdot 10^{-5}$, TSKHAI [761] $P_{\Sigma M} = C \cdot 5 \cdot 9{\cdot}10^{-5}$, FISCHER and KROPP [262] $P_{\Sigma M} = C \cdot 1{\cdot}2 \cdot 10^{-4}$, $P_{\Sigma M}$ being the partial pressure in atmospheres and C the molar concentration of the element in the solution. Thus, we may calculate with a mean value of

$$P_{\Sigma M} = C \,.\, 10^{-4} \tag{12.10}$$

When dispersion devices with direct injection are used, the proportionality constant is roughly one order greater [261], which influences the manifestation of the individual types of interference.

It is seen that in the acetylene–air flame the degree of ionisation reaches 10 % for lithium at a molar concentration of $3 \cdot 10^{-4}$, for sodium at 10^{-3}, for potassium at $3 \cdot 10^{-2}$, for rubidium at 10^{-1}, and for caesium, at $3 \cdot 10^{-1}$. With larger concentrations the degree of ionisation is less, with lower concentrations the degree of ionisation increases. The concentration of atoms in the flame, and therefore also the intensity of their radiation, increases more rapidly than their concentration in the solution, with the result that the calibration graphs are curved. This is a well-known phenomenon, which occurs in the analysis of K, Rb, and Cs.

If the relation between the concentration of the element in the solution and its partial pressure in the flame is known, it is possible – with an accuracy corresponding to the validity of the above relation – to estimate the degree to which the element analysed is dissociated or ionised. Assuming that $P_M = P_X$, or that $P_{M^+} = P_{e^-}$, modification of the above expression for the equilibrium constant shows that the degree of dissociation $\alpha_{dis} = P_M/P_{\Sigma M}$, or the degree of ionisation $\alpha_{ion} = P_{M^+}/P_{\Sigma M}$ are only functions of the overall concentration of the element in the flame and of the respective equilibrium constant according to the relation [591]

$$\alpha = \frac{\sqrt{4a + 1} - 1}{2a}, \tag{12.11}$$

where $a = P_{\Sigma M}/K$.

This relationship is plotted in Fig. 12.13, from which the degree of dissociation or ionisation may be read if the respective dissociation or ionisation constant and partial pressure in the flame are known.

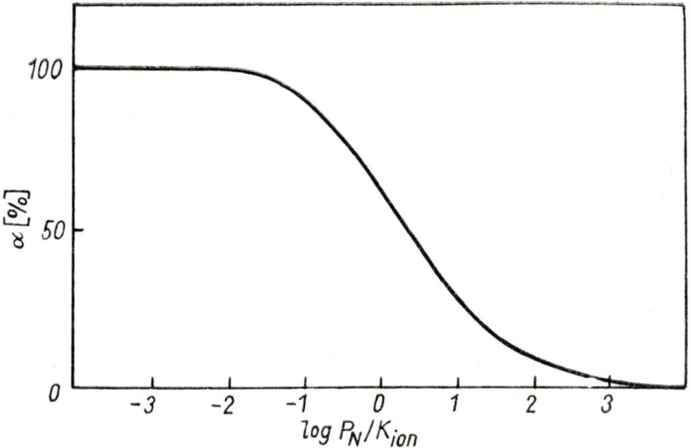

Fig. 12.13. Dependence of the degree of ionisation on the element concentration in the flame

P_N — partial pressure of free atoms, K_{ion} — ionisation constant

It may easily be proved by means of the equilibrium constant values, of Fig. 12.13 and relation 12.10, that when 10^{-5} to 10^{-1} molar solutions are dispersed, ionisation is practically unimportant in relatively cool flames, while in hotter flames, e.g. acetylene – air, dissociation of halides is practically complete. Having this knowledge it is now possible to interpret some observed interferences.

When the anion concentration in the solution rises, e.g. due to the addition of excess acid, the signal of the element in the flame decreases. The mechanism, however, differs in the town gas and acetylene flames.

In the first case, the dissociation equilibrium shifts and the free atom concentration decreases. In the acetylene – air flame, where dissociation is complete, the electronegativity of anion-forming elements becomes effective and the ionisation equilibrium is shifted at the expense of free atoms. According to POLUEKTOV [592], the shift of equilibrium is caused by a loss of metal ions according to relation 12.12. This statement is based on the observation, that the conductivity of the flame, which is mainly determined by the electron concentration, is not decreased by the addition of the acid

$$M^+ + X = MX^+ \tag{12.12}$$

According to ZHITKEVICH, et al. [846], electrons are lost in agreement with the relation

$$e^- + X = X^- \tag{12.13}$$

The accuracy of this latter relation is confirmed by the rise of intensity due to an addition of halides, observed in the case of the ionic lines of alkali earth elements, as well as by the decreased amount of electrons in the flame as determined by means of a microwave method [847].

The anion addition therefore causes a shift of the ionisation equilibrium in the case of hotter flames. There is an exception in the case of the nitrate ion since when this ion decomposes, no other products capable of shifting the existing equilibrium in the flame are formed. The equilibrium shift is more probable in apparatus equipped with direct injection of the aerosol into the flame compared to equipment with a mist chamber [261]. In the latter case, however, the acid addition usually also influences the transport effects, i.e. the dispersion process, size of droplets, and aerosol coagulation in the mist chamber. From this it follows that the interference which is usually called the anion effect, may be due to many different mechanisms.

From SAHA's equation it follows that ionisation is substantially weaker at lower temperatures. In some cases, therefore, this interference may be avoided by the use of a cooler flame. For example, in the town gas – air flame the mutual influence of alkalis is negligible [212]. On the other hand, when a flame of lower temperature is used, the concentration of excited atoms decreases also, and with it the intensity of the emitted line. To increase the sensitivity of the determination of alkali metals, a flame of higher temperature must therefore be used, and ionisation must be suppressed by the addition of an element easy to ionise.

The influence of electronegative elements is eliminated easily if the concentration of acids is kept constant in the samples as well as standards. It is more difficult to remove the mutual influence of alkali metals. Although the individual alkali metals occur in specific concentration ratios in certain types of samples, usually their concentration varies to such an extent that the error involved is excessive. In this case, there are two principle possibilities. Either the amount of the interfering alkali metals is determined and a correction is applied, or one of the alkali metals is added in such an amount that the original concentration becomes negligible or that the final concentration is constant in all samples. Thus the resulting ionisation equilibrium in the flame is normalised to a certain value. It is recommended to add caesium in the role of this normaliser [261], [689], which, due to its low ionisation potential, has the greatest effect, so that it suppresses the ionisation of potassium and rubidium too. Unfortunately, caesium salts of sufficient purity are too expensive for routine analysis. To determine rubidium and caesium, additions of potassium salts are frequently used [249–251], [698]. Here the disadvantage is that potassium, having a high ionisation potential, must be added in a great excess (25 to 30 g K_2SO_4 per litre), causing some secondary difficulties. Conventional dispersion devices with indirect injection do not permit solutions of such large concentrations to be dispersed, and the nozzles are easily clogged. FUKUSHIMA [286] therefore recommends the use of two dispersion devices in parallel, one of which is used for the solution analysed and the other for the concentrated potassium chloride solution. A second difficulty involves the fact, that the intensity of the potassium doublet is very great, increasing the demands on the optical system and dispersion of the instrument, so that with the majority of commercial instruments rubidium and caesium cannot be determined in this case due to radiation interference. The procedure may also be modified by first determining the potassium content in the sample and arranging the final concentration in the sample, as well as in standard solutions, to a certain level. Although this does not suppress the ionisation of rubidium and cesium completely, it is at least guaranteed that no substantial shift of the ionisation equilibria in the flame can take place.

When light alkali metals are being determined – lithium and sodium – the potassium concentration needed to suppress ionisation becomes substantially less, and in these cases potassium is very widely used as "normaliser". For the determination of potassium this method is rather seldom used. Rubidium and caesium salts are very expensive, as already mentioned, and the required amount of sodium is too great: lithium is practically unsuitable for this purpose. As far, of course, as the potassium content is greater than that of sodium, the influence of sodium may be neglected. In the reverse case, the method of the system of calibration curves is chosen, as described in section 10.7.2. If the potassium content also varies to a greater extent, its reverse influence on sodium must also be taken into account in high-accuracy analyses.

The additions must be selected in such a way that the number of free electrons formed by the ionisation of interfering elements should be negligible compared to the number of electrons obtained from the normaliser. From the expression for the ionisation constants of elements, the following expression may be derived for the final concentration of electrons formed by the ionisation of the elements M_1, M_2, M_3

$$P_{e^-} = P_{M_1}K_1 + P_{M_2}K_2 + P_{M_3}K_3 \dots \qquad (12.14)$$

Hence it follows that the normaliser addition must be chosen so as to make the product of its ionisation constant and partial pressure sufficiently greater than similar products of the interfering elements [543]. It is evident therefore that a smaller amount need be added if an easily ionised element, i.e. with a greater ionisation constant value, is used. Therefore, caesium is recommended as normaliser.

12.4 Radiation interferences and their elimination

Another group of interferences are the radiant or radiation interferences. In emission flame photometry this group is understood to include all radiation falling onto the detector which does not belong to the element analysed. This radiation may include light of various wavelengths dispersed in the optical system of the instrument, and furthermore radiation of a wavelength which belongs in the spectral interval isolated by the instrument and caused by other, non-analysed components of the solution (i.e. at the given wavelength) or of the flame. Dispersed radiation participates mainly if the element analysed has an intense radiation (e.g. sodium) and the spectral sensitivity distribution of the detector is unfavourable to the element analysed, i.e. when the lines measured lie in the extremes of the spectral sensitivity of the detector, in the ultraviolet region below 350 nm or in the infrared region above 700 nm. With instruments of high quality, however, the amount of dispersed radiation should be small to such an extent that this interference should be manifested only with an excess of several orders of magnitude of the interfering component. When required, this interference can be eliminated easily. Coloured filters are usually placed in the optical path to keep back the interfering radiation. For example, to suppress the influence of yellow sodium radiation a didymium filter is used. For potassium the respective interference filter may be used as a mirror to reflect selectively the radiation of other wavelengths.

A greater danger to the accuracy of the analysis is involved in radiation the wavelength of which lies within the spectral interval isolated by the filter or optical dispersion system of the instrument. The magnitude of this interference obviously also depends on the width of this interval, being far greater in the case of filter-type instruments than with monochromators. With the latter, the slit width is another important factor.

The magnitude of the radiation interference of an element, for a specific line and experimental conditions may be expressed easily in terms of the specificity factor [177], [587]. In principle, this is a quantity which states the excess of the interfering element which, under the conditions of measurement of the element analysed, causes the same deflection as the element analysed. It is generally calculated from the relation

$$F_{x,y} = \frac{I_x \cdot c_y}{I_y \cdot c_x}, \qquad (12.15)$$

where I_x is the galvanometer deflection at a concentration of the element analysed, c_x, and I_y is the galvanometer deflection at a concentration c_y of the interfering element. With this relation, it is unimportant whether a coinciding line or molecular band is involved, or a true continuous background, or a background caused by dispersed radiation as already discussed earlier. It is only assumed that the deflection is proportional to the concentration of the respective element. The specificity factor then characterises not only the mutual influence of the element, but also the experimental conditions. Obviously the radiation influence may be eliminated by subtracting the deflection which corresponds to the parasitic radiation. As long as the concentration of the interfering elements in the samples is known, model solutions can be prepared and the deflection corresponding to the blank experiment can be measured.

In some cases, interfering influences may be decreased by suppressing the radiation of the interfering component. For example, in the determination of lithium, the molecular band of SrOH interferes; in the determination of sodium it is the CaOH band. As mentioned earlier, the addition of some components to the analysed solution may be used to bind alkaline earth elements in low-volatile compounds and thus to decrease their radiation intensity. The influence of strontium was eliminated by adding ammonium phosphate [655]. SCHUHKNECHT and SCHINKEL [685] recommend an addition of aluminium salts to eliminate the radiation interference of calcium in the determination of sodium.

When a monochromator is used, the intensity of the interference background may be measured on both sides of the measured line, the mean value being subtracted from the measured intensity. The procedure is illustrated in Fig. 12.14. When a recording instrument is available, it is usually quicker and more accurate to read these values off the record.

With constantly rising demands on the sensitivity of analyses, corrections for background radiation are becoming essential. To this purpose, instruments were designed to record the spectrum automatically in a narrow wavelength interval: this greatly facilitates the correction procedure [589], [788]. DAVIS, et al. [169], described the design of an integrated flame photometer for the determination of strontium traces in calcium salts: the basic unit is a Uvispek monochromator made by the Hilger Co. This instrument not only permits corrections for the background, it can also be used for measurement by the internal standard method. Thus it achieves a high degree of precision and accuracy. To determine sodium in limestone, PLŠKO [579] modified the Model ISP 51 spectrograph with a camera of $f = 270$ to give a flame spectrophotometer with simultaneous recording of two lines. This allows the internal standard method to be used equally with the lines located in a region with a strongly variable background.

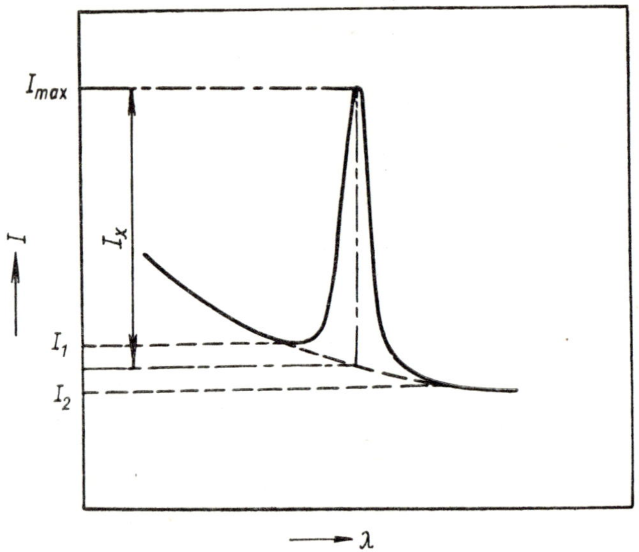

Fig. 12.14. Correction for a variable background $I_x = I_{max} - \frac{I_1 + I_2}{2}$

Differing from the emission method, analogous interference effects are very improbable in the absorption method, since the spectral interval measured is far narrower – it is defined by the width of the line irradiated. Some molecular spectra may have an interfering effect. It is stated that alcohol has a strong absorption spectrum below 230 nm, so that it interferes in the determination of zinc at 213.8 nm and of Cd at 228.8 nm [118]. Similarly, ROBINSON [645] observed interfering absorption by individual components of the OH band in the oxygen – hydrogen flame in the case of the lines Pb 280.2, 283.3 and 317.5, Pt 283.0, 306.5, Cu 282.4 and Ta 310.3 nm. Other lines of these elements, even though located in the same spectral region, are not affected by this interference. Some authors [310] have stated that the presence of hydrochloric acid also increases the absorption of zinc, ascribing this influence to the absorption of the hydrogen chloride molecular spectrum. It was later found, however, that in this case an increase of the concentration of zinc atoms in the flame was involved, caused by more rapid liberation of zinc atoms with rising acidity of the solution. Therefore, some degree of care is required when explaining the observed phenomena.

The molecular bands of alkaline earth elements also ought to exhibit an interfering influence. In reality, however, this does not occur. In the case of atomic absorption flame photometry, radiation interferences like the one of potassium in the determination of rubidium, do not occur. On the other hand, there are some interferences which have no analogue in the emission method. For example, if the interfering element emits a line or band in the flame in the spectral interval limited by the monochromator or filter, the resulting radiation incident on the detector is increased. This interference may easily be eliminated by using a modulated source of light and an alternating amplifier, tuned to the respective frequency. Details of this case will be found in section 9.2.

Another type of interference may be encountered in the determination of transition elements (Mn, Fe, Cr, Ni) which have very rich spectra. In this case, the discharge lamp employed may emit other lines also in the same spectral interval, and these need not necessarily be resonance lines. These lines, however, are not absorbed in the flame and they cause the working curves to bend [19]. Such interference obviously influences the standards as well as the samples, decreasing the precision of the analysis only, as it does not cause a systematic error.

In conclusion it may be stated that with the absorption method there are substantially fewer interfering influences and they are weaker than with the emission method. Thus there are fewer causes of error in determinations carried out by means of atomic absorption.

12.5 Buffer solutions in flame photometry

In a number of papers it is suggested that buffer solutions be added to the sample as well as to standard solutions in order to suppress the interfering influences. To a large degree, the resulting effect is similar to the effect of buffers in classical chemical analysis, i.e. concentration variations of the interfering component in the sample are not manifested in the presence of a buffer. From a strict point of view, the term "buffer" should only be applied to substances which in some way stabilise the reaction equilibrium in the flame. For example, excess acid stabilises dissociation equilibria, an element easily ionised stabilises the ionisation equilibria. Frequently, however, the word buffer is extended to include releasing elements, which suppress the formation of non-volatile compounds of the element analysed, or such elements that stabilise and enhance the background spectrum (e.g. potassium in the determination of strontium [383]), as well as other reagents of different functions. In flame photometry the term buffer is conventionally a very broad one, denoting all the additive solutions by means of which we attempt to improve the accuracy of the measured results.

12.6 The influence of organic solvents

Organic reagents are sometimes employed to increase the sensitivity of a determination. The effect of organic solvents was already mentioned in the discussion of the individual interfering influences, but we prefer to present the problem once more in a comprehensive manner since a widely employed technique is involved.

Many studies have been devoted to this problem [104], [214], [216], [618], [620], [642], [667]. It follows from a comparision of the individual authors'conclusions that this influence has different explanations with different experimental conditions, instruments, elements, and lines. Organic solvents doubtless influence transport effects, i.e. processes which take place on dispersion of the solution, fineness of the aerosol, and finally also the amount of the solution which enters the flame. This influence is more marked in the case of instruments fitted with a mist chamber, where coagulation and sedimentation decrease the overall "aerosol yield". At the same time, the influence of organic solvents is more distinct in the case of imperfect or low efficiency dispersion devices compared with more efficient ones. In the case of some elements which have no tendency whatever to form compounds in the flame, e.g. silver, zinc, or to some extent, copper, the increase of the overall signal may be explained by the increased amount of the element which enters the flame within a unit of time. For example, ALLAN [23] found that the ratio of the amounts of copper passing through the burner before and after the use of an organic solvent corresponds virtually exactly to the observed ratio of signals in the two cases. Similarly, ELWELL and GIDLEY [230] increased the amount of aerosol entering the flame 1.8 times by an addition of 20% isopropyl alcohol, and similarly the signals for zinc increased 1.8 times, for lead 1.9 times, for magnesium 2 times and for iron 2.4 times. In the case of iron and magnesium some other interfering effect evidently also participated.

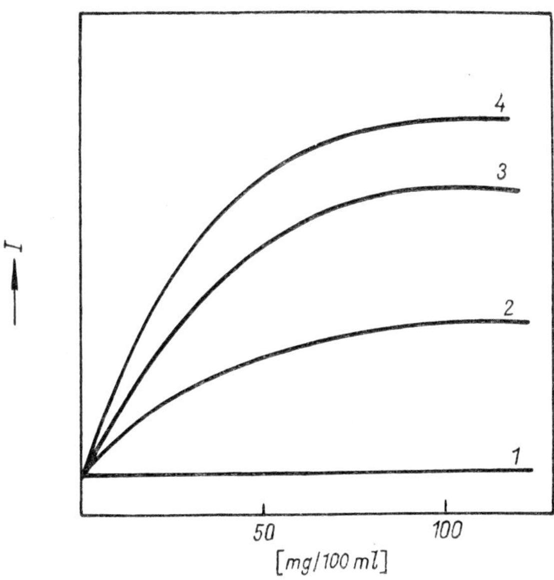

Fig. 12.15. Influence of rising content of Na, Rb and Cs on the emission intensity of potassium

I — intensity, c — concentration, 1 — intensity of potassium emission, 2 — in the presence of Na, 3 — in the presence of Rb, 4 — in the presence of Cs

Besides transport processes, the organic solvent also influences the flame temperature. In the case of direct-injection burners, dispersed water substantially decreases the flame temperature, which circumstance is called the cooling effect of water. The intensity of the radiation studied in this case depends on the amount of aqueous solution which enters the flame in a unit of time. With rising amount of the solution introduced, the concentration in the flame of atoms of the element to be determined rises, but due to the temperature decrease the amount of excited atoms decreases in accordance with the Boltzmann equation (equation 5.12). The result of these two opposite processes is rather a sharp intensity peak of the individual lines in relation to the amount of solution introduced into the flame [39], [278], [287]. When organic solvents with a substantially lower heat of evaporation decrease the cooling effect, the flame temperature rises and substantial changes take place in the position and absolute value of the maximum emission. In the case of instruments fitted with a mist chamber, where the actual amount of solution entering the flame is relatively small, the cooling effect is unimportant. Nonetheless, organic solvents influence the flame temperature in the upward or downward direction, depending on the composition of the fuel mixture, although the extent of this influence is substantially less than in the preceding case [36]. A typical relationship of the temperature of an acetylene – air flame with dispersion of acetone and water solutions is illustrated in Fig. 12.16. Generally, the intensity of the emitted

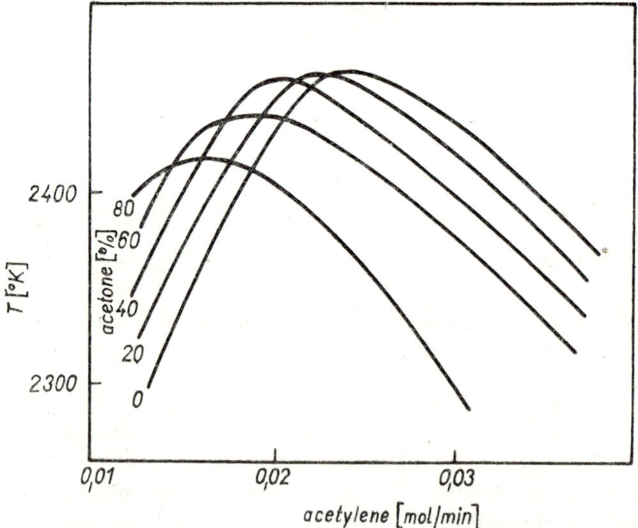

Fig. 12.16. Influence of acetone content in the dispersed solution on the temperature of the acetylene – air flame with constant flow rate of air. Burner with mist chamber

and absorbed lines is increased at the same time, indicating an increased concentration of free atoms and not a simple rise of the number of excited atoms. This means that the rate of dissociation of the atom from various complexes and compounds in the flame is influenced too. Some elements may form organic complexes with the solvent. On dissociation of these complexes in the flame, free atoms are then formed substantially more rapidly than from hydrated ions originating in an aqueous solution. When elements are isolated by means of extraction of organic complexes into an organic solvent, it is possible to determine by flame-photometric means those elements also which form very stable oxides. For example, titanium has been determined after extraction of its oxinate into chloroform [130], and aluminium was determined after extraction of the cupferronate into 4-methyl-pentanone-2 [204]; similar methods were applied to the study of scandium, yttrium, and rare earth spectra [622], vanadium, niobium, rhenium, molybdenum, tungsten, and titanium [254], and of the absorption spectra of vanadium, titanium, niobium, scandium, and rare earths [253], aliminium, beryllium and barium [691]. In all these cases, burners with direct injection of the aerosol into the flame and with relatively hot oxygen – hydrogen or acetylene – oxygen flames were employed for emission as well as absorption determinations.

Even under these conditions, the emission of the elements mentioned is limited to a very narrow region, and therefore much depends on the correct positioning of the burner when these spectra are being studied. Emission or absorption of these elements is limited, more or less, to the reaction zone of the flame or its immediate vicinity, as far as this part of the flame can be distinguished sufficiently exactly with direct-injection burners. In some cases, the intensity of the emitted radiation is evidently enhanced by chemiluminescence [309], [318], [848], e.g. in the case of tin.

It is to be seen from this brief review, that organic solvents have a far-reaching influence on all processes in flame photometry: in the case of transport processes the influence of solvents is non-specific, with processes taking place in the flame, however, the situation is clearly differentiated. Elements which have a tendency to form oxides are generally influenced more than those which tend to dissociate.

We have attempted in this chapter to present a survey of the individual interfering influences which occur in flame photometry, and to illustrate

them by means of typical examples from practical analysis. It must be pointed out that in the prevalent majority of cases the individual interferences do not occur isolated in the way they were described. Generally, several interfering influences overlap, and which of them will be manifested more distinctly depends, to a large extent, on the specific conditions. This circumstance of course makes it rather difficult to elucidate the true causes of erroneous results and to eliminate them. These problems must be solved individually by every analyst who is concerned with working out new methods, and the preceding discussion can be but a guide-line to this purpose.

With the individual interfering influences we have also indicated the methods of their elimination. We did not mention any general procedures for the elimination of interfering influences, e.g. different methods of additions or other procedures of evaluation, as these have been discussed in section 10.12 already.

13. **Emission and absorption spectra of elements**

13.1 **General properties**

It was thought up to recent times that the application of flame photometry would continue to be limited to the determination of rather a small number of elements. Increasingly perfect instrumentation, however, has made it possible in recent years to measure the radiation emitted in flames by nearly all the elements, while it has become possible to determine other elements by means of absorption flame photometry. Theoretically, therefore, flame photometric means could be used to determine all elements. Although practically this ideal state has not yet been achieved by any means the development of flame photometry is certainly not yet concluded, and it may be expected that this technique will find a substantially wider field of application in practical analysis. Nowadays, emission flame photometry is being used (with all the advantages inherent in this method) to determine alkali metals, alkaline earths, indium, gallium, thallium, copper, and, occasionally, some other elements also. The absorption technique widens the possibility of advantageous determination to many other elements, the most important of which are magnesium, zinc, cadmium, mercury, copper, silver, gold, platinum group metals, etc.

There are several reasons for which the emission flame photometric technique was up to now limited to the elements mentioned. Firstly, flames of rather high temperatures must be used for the analysis of other elements, since only thus can thermally stable molecules be dissociated sufficiently, and some of the elements difficult to excite can likewise be excited at elevated temperatures only. Under such conditions, however, the selectivity of the determination is decreased, since the spectra emitted are composed of many lines, so that the possibility of coincidences is substantially greater. At the same time, there are difficulties related to the increased background. Obviously therefore, isolation and sometimes also detection of the radiation studied is more difficult. Work with gases which form flames of such high temperatures is inconvenient, and far more complicated than in the case of conventional flames: more complicated instrumentation is likewise required. It is true that apparatus fulfilling these demands has already been designed and described, but these instruments are not yet manufactured commercially. It will be easier to utilise chemiluminescence when working with some organic solvents or strongly reducing flames. It was observed that this effect contributes to a substantial increase of the radiation intensity of Sn, Ge, Pb, As, Sb, Bi, Te, and of a number of other elements [104], [318].

In absorption flame photometry the situation is similar, although excitation of the atoms is not required in this case. Again, the limiting factor is the lifetime of free atoms in the flame. Moreover, there is the need to design a light source to emit the spectrum of the element in question with not only sufficient intensity, but also stability. For practical purposes this source must also have rather a long lifetime. These conditions are not fulfilled for all elements, and therefore the practical application of absorption flame photometry likewise continues to be limited to a certain group of elements.

With both methods, the literature frequently includes data concerning the sensitivity with which a specified element can be determined at a certain wavelength. These data are relative and they cannot be simply transferred from one case to another. The sensitivity of a determination depends on the flame employed, its temperature, and the reactions which take place in the flame, the solvent employed, sample composition, type and efficiency of the dispersion device, properties of the filter or monochromator, and finally, also the properties of the detector. These relationships are very complicated and the experimental data published by different authors may differ by several orders of magnitude. A theoretical sensitivity calculation is only possible for specific conditions, for a specified element, and under simplifying assumptions [821].

When the possibilities of determining elements by means of measurement of line spectra are to be assessed, the basic condition is knowledge of the degree of dissociation of the element involved, which influences the final concentration of free atoms in the flame; furthermore the oscillator strength of the respective line, which determines the probability of a given transition, and in the emission method also the excitation potential of the line, which determines the number of excited atoms. The oscillator strength values of some analytical lines and their excitation potentials are given in Table 13.1. The degree of dissociation of an element in the flame obviously also depends on the composition of combustion products: it will be described qualitatively for the individual elements in the following text.

For a rough but rapid orientation in the possibilities of emission flame photometry in the determination of individual elements, Table 13.2 lists the main lines employed and the molecular spectrum bands. Less important lines and bands are not included, even though they may interfere in the determination of other elements, as far as there is an excess of the interfering element. The second column of the table states the particle to which the radiation in question is ascribed, the third states the wavelength of the lines or peak of the molecular bands, and the fourth gives the limit of detection or sensitivity as stated by different authors: since each author defines the values in a different way, the data vary to some degree. A unified manner of expressing these two quantities has not yet become conventional in the literature: therefore we are presenting the data as they were available to us. The last column of the table includes those elements whose presence in the sample may cause radiation interference. The table, however, lists only those interferences which are caused by conventional elements, or which may occur even in the

Table 13.1. EXCITATION ENERGY AND OSCILLATOR STRENGTH OF SOME LINES

Line	nm	eV	f	Line	nm	eV	f
Li I	670.8	1.9	0.71	Tl I	276.8	4.44	2.27
Na I	589.0	2.11	0.76		377.6	3.28	0.13
	589.6	2.0	0.39		535.0	3.28	0.14
K I	766.5	1.62	0.69	Sn I	286.3	4.32	0.65
	769.9	1.61	0.34	Pb I	217.0	5.67	0.048
Rb I	780.0	1.59	0.80		283.3	4.40	0.22
	794.8	1.56	0.40		405.8	4.38	0.46
Cs I	852.1	1.46	0.80	Sb I	206.8	5.98	0.10
	894.4	1.39	0.40		217.6	5.69	0.045
Mg I	285.2	4.34	1.2		231.1	5.36	0.03
Ca I	422.7	2.93	1.49	Bi I	206.2	6.00	0.095
Sr I	460.7	2.69	1.54		222.8	5.55	0.0025
Sr II	407.8	3.04	0.76		223.1	5.55	0.012
Ba I	533.6	2.24	1.40		306.8	4.04	0.25
Ba II	493.4	2.51	0.33	Cr I	357.9	3.46	0.34
Cu I	324.8	3.82	0.74		425.4	2.91	0.1
	327.4	3.78	0.38	Mo I	313.3	3.96	0.20
Ag I	328.1	3.78	0.53		317.0	3.91	0.12
	338.3	3.66	0.23		390.3	3.17	0.67
Au I	242.8	5.10	0.16	Mn I	276.5	4.43	0.58
	267.6	4.63	0.12		403.1	3.08	0.056
Zn I	213.9	5.80	1.2		403.3	3.08	0.040
	307.6	4.03	0.000 17		403.4	3.08	0.027
Cd I	228.8	5.41	1.2	Fe I	248.3	4.99	0.34
	326.1	3.80	0.001 8		252.3	4.91	0.30
Hg I	184.9	6.67	1.2		271.9	4.16	0.15
	253.7	4.88	0.03		296.7	4.18	0.056
Al I	396.2	3.14	0.15		302.1	4.11	0.083
Ga I	287.4	4.29	0.32		372.0	3.32	0.057
	403.3	3.07	0.13		386.0	3.21	0.034
	417.2	3.07	0.14	Co I	240.7	5.15	0.22
In I	303.9	4.10	0.36		242.5	5.11	0.19
	410.2	3.02	0.14		345.4	4.02	0.46
	451.1	3.02	0.16		352.7	3.51	0.048
				Ni I	231.1	5.36	0.066
					232.0	5.34	0.095
					341.5	3.65	0.14
					352.5	4.54	0.12
				Pd I	244.8	5.06	0.074
					247.6	5.00	0.10
					340.5	4.46	0.29
					363.5	4.23	0.17

Table 13.2. EMISSION SPECTRA OF ELEMENTS

Element	Specification of radiation	Wavelength nm	Character of radiation	Excitation energy eV	Limit of detection mg/1 [373]	Limit of detection μg/ml [255]	Emission sensitivity p. p. m. / % T [177]	Coinciding radiation
Ag	Ag	328.1	l	3.78	0.2	0.3	1.0[1]	Na, Cu, OH
	Ag	338.3	l	3.66	0.2		0.6[1]	(Na, OH)
Al	Al	394.4	l			0.3		
	AlO	464	p					(Nm)
	AlO	467	p		65			(In)
	AlO	484	p		35		0.4	
Au	Au	267.6	l	4.63	6	15	(0.0)[1]	
B	BO	494	p		2			
	BO_2	518	p		1.7		3	Ba
	BO_2	548	p		0.7		3	Ba, Ca
Ba	Ba	493.4	i	5.21; 2.51	2		4	
	BaOH	513	p		3		1[1]	
	Ba	553.6	l	2.24	1	0.05	1[1]	Ca
	BaOH	830	p		0.7			
Ca	Ca	422.7	l	2.93	0.04	0.005	0.07	
	CaOH	554	p		0.06		0.16	Ba
	CaOH	622	p		0.02		0.6	Na, Sr, (Li)
Cd	Cd	228.8	l	5.41	40	9	10[1]	
	Cd	326.1	f	3.80	40	30	0.5[2]	
Co	Co	345.4	d	4.02	1.5	1	3.4	

Table 13.2 — Continued

Element	Specification of radiation	Wavelength nm	Character of radiation	Excitation energy eV	Limit of detection mg/l [373]	Limit of detection μg/ml [255]	Emission sensitivity p. p. m./% T [177]	Coinciding radiation
		345.5		3.82				
	Co	350.2	d	3.97	3			
		350.6		4.05				
		352.7		3.51	1.5		4.0	
	Co	352.9	t	3.68				
		353.0		4.03				
Cr	Cr	360.5	l	3.43		0.1		
	Cr	425.4	l	2.91	0.5			
	Cr	427.5	l	2.90	0.6		20	
		520.5		3.32				
	Cr	520.6	t	3.32	1			
		520.8		3.32				
Cs	Cs	455.5	l	2.72	30	8	2.0	In
	Cs	459.3	l	2.70			8.0	(In)
	Cs	852.1	l	1.46	0.01	0.008	0.5[1]	
	Cs	894.4	l	1.39	0.01		0.5[1]	
Cu	Cu	324.8	l	3.82	0.1	0.1	0.6	(OH)
	Cu	327.4	l	3.78	0.1	0.2	0.8	(OH), Na, Ag
	CuOH	537	p		1.7			Tl
Dy	DyO	549	p					Ba, Ca

Table 13.2 — Continued

Element	Specification of radiation	Wavelength nm	Character of radiation	Excitation energy eV	Limit of detection mg/1 [373]	Limit of detection μg/ml [255]	Emission sensitivity p. p. m. /% T [177]	Coinciding radiation
	DyO	573	p		0.08[1]		0.08[1]	(Na)
Er	ErO	552	p		0.12			Ba, Ca
Eu	Eu	459.4	l	2.70	1	0.003	0.051	In
	(EuOH)	598	pk		1			Na, Ca, Sr
Fe	Fe	372.0	l	3.32	0.7	0.7	2.5	Tl, Mg
	Fe	386.0	l	3.21	0.9		2.7	Mg
Ga	Ga	403.3	l	3.07	1		1	Mn, K
	Ga	417.2	l	3.07	0.5	0.07	0.5	Ca
Gd	GdO	426	p		0.6[1]		0.1[1]	
	GdO	570	p				0.14[1]	
Hg	Hg	253.7	f	4.88	6[2]	40	(10)[2]	
Ho	HoO	532	p		0.2[1]			Tl
	HoO	566	p		0.08[1]			
In	In	410.2	l	3.02	2.0		0.14[1]	Ga
	In	451.1	l	3.02	0.15	3.03	0.07[1]	
K	K	404.4	d	3.06	1		1.7	Ga, Mn
		404.7		3.06				
	K	766.5	l	1.62	0.0003	0.003	0.02	
	K	769.9	l	1.61	0.0005	0.02		Rb
La	LaO	438	p		0.6		0.7	

Table 13.2 — Continued

Element	Specification of radiation	Wavelength nm	Character of radiation	Excitation energy eV	Limit of detection mg/l [373]	Limit of detection μg/ml [255]	Emission sensitivity p. p. m. / % T [177]	Coinciding radiation
	LaO	442	p		0.6		0.7	
	LaO	560	p		0.6		0.6	Ba, Ca
Li	Li	610.4	l	3.87	1	0.1	4.4	Na, Sr, Ca
	Li	670.8	l	1.90	0.001	0.001	0.067	Sr
Lu	LuO	466	p		0.3[1]		0.05[1]	
	LuO	517	p		0.2[1]			Ba
Mg	Mg	285.2	l	4.34	0.006	0.2	1.0	
	MgOH	370	p		2		1.4[1]	Tl
	MgOH	381	p		3		1.6[1]	
Mn		403.1			0.02	0.1	0.1	
	Mn	403.3	t	3.08				Ga, K
		403.5						
	MnO	539	p		1.2			Tl
	MnO	559	p				2.7	Tl, Ba, Ca
	MnO	588	p		0.6			Na, Ca
Na	Na	589.6	d	2.10	0.0004	0.0005	0.001[1]	Ca
		589		2.11				
	Na	330.3	d	3.75	1			Cu
	Na	819	d	3.61	0.5			(Ba)
Nd	NdO	599	pk					Na, Sr, Ca

Table 13.2 — Continued

Element	Specification of radiation	Wavelength nm	Character of radiation	Excitation energy eV	Limit of detection mg/1 [373]	Limit of detection μg/ml [255]	Emission sensitivity p. p. m./% T [177]	Coinciding radiation
	NdO	702	pk		1		1[1]	
	NdO	712	pk		1			
Ni	Ni	339.3	d	3.67				
	Ni	341.5	l	3.65	1	1	3.5	
	Ni	349.1	l	3.65	1.4			
	Ni	351.5	l	3.63	1.1			
	Ni	352.5	l	3.54	0.7	0.6	1.6	
	Ni	361.9	l	3.85	2			Mg
Pb	Pb	368.4	l	4.34	20	3	21	
	Pb	405.8	l	4.38	15		14	K, Ga, Mn
Pd	Pd	340.5	l	4.46	1		(0.15)[1]	Ag
	Pd	363.5	l	4.23	1		(0.1)[1]	Mg
Pr	PrO	569	p					Ba, Ca, (Tl)
	PrO	576	p		0.7[1]			Tl, Na
Rb	Rb	780	l	1.59	0.005	0.002	0.6[1]	K
	Rb	794.8	l	1.56	0.006			(K)
Rh	Rh	343.5	l	3.60			(2.0)[1]	
	Rh	369.2	l	3.35	0.7	0.3	(0.7)[1]	Mg, (Tl)
Ru	Ru	372.8	l	3.32	0.3	0.3	(0.3)	Tl, Mg

Table 13.2 — Continued

Element	Specification of radiation	Wavelength nm	Character of radiation	Excitation energy eV	Limit of detection mg/1 [373]	Limit of detection μg/ml [255]	Emission sensitivity p. p. m./% T [177]	Coinciding radiation
Sc	ScO	604	p				0.012[1]	Na, Sr, Ca
	ScO	607	p		0.3			Na, Sr, Ca
Sm	SmO	624	pk		4		0.25[1]	Ca
	SmO	652	pk		3		0.25[1]	
Sr	Sr	460.7	l	2.69	0.015	0.004	0.6	
	SrOH	605	p		0.01[1]		0.28	Na, Cr
	SrOH	666	p		0.015[1]			Li
Tb	TbO	534	p		0.25[1]		0.14[1]	Ba, Ca
	TbO	592	pk		0.12[1]			Na, Ca
	TbO	598	pk		0.15[1]			Na, Ca
Ti	TiO	517	pk		7		10[1]	Ba
	TiO	545	pk		7			Ba, Ca
	TiO	715	pk		1			
Tl	Tl	377.8	l	3.28	1	0.09	0.6[1]	Mg
	Tl	353.1	l	3.28	2		1.2[1]	Ba
Tm	TmO	491	p		0.35[1]			
V	VO	523	p		3.5		12[1]	Ba, (Ca)
	VO	547	p		2.5			Ba, Ca
	VO	574	p		1.7			
Y	YO	482	p		0.31[1]			

Table 13.2 — Continued

Element	Specification of radiation	Wavelength nm	Character of radiation	Excitation energy eV	Limit of detection mg/1 [373]	Limit of detection μg/ml [255]	Emission sensitivity p. p. m./% T [177]	Coinciding radiation
Y	YO	599	p		0.3		0.2	Na, Sr, Ca
	YO	615	p		0.3		0.4	Na, Sr, Ca
Yb	(YbOH)	478	p					
	(YbOH)	485	p					
	(YbOH)	498	p		(2)			
	(YbOH)	533	p		(2)			Ba, Ca
	(YbOH)	555	p		(1)		0.06[1]	Ba, Ca
	(YbOH)	573	p		(1)			
Zn	Zn	513.9	l	5.8	600	50	500	

l — line corresponding to a neutral atom, d — doublet, t — triplet, f — intercombination line, due to transition between terms of different multiplets, i — ionic line, p — molecular band, k — continuum, caused by a very broad non-differentiated band: the wavelength stated lies within the maximum intensity region

( ) — uncertain data or circumstances which occur only exceptionally, [1] — hydrogen—oxygen flame, [2] — hydrogen-air flame.

Unless mentioned otherwise, data refer to the acetylene-oxygen flame.

Table 13.3. SPECTRA OF ELEMENTS FOR ABSORPTION FLAME PHOTOMETRY

Element	Line nm	Sensitivity p.p.m.	Limit of detection p.p.m.	Gas mixture
Ag	328.1	0.05	0.01	C_2H_2—air
	338.3	0.15		C_2H_2—air
Al	309.3	1.1	0.1	N_2O—C_2H_2
		2.0	0.5	C_2H_2—O_2
As	193.7	5.0	0.5	C_2H_2—air
Au	242.8	0.3	0.1	C_2H_2—air
	267.6	1.3		C_2H_2—air
B	249.7	35	6	N_2O—C_2H_2
Ba	553.6	0.4	0.1	N_2O—C_2H_2
		5	1	C_2H_2—air
Be	234.9	0.03	0.003	N_2O—C_2H_2
Bi	223.1	0.68	0.02	C_2H_2—air
	222.9	1.5		C_2H_2—air
	306.8	2.1		C_2H_2—air
Ca	422.7	0.08	0.003	C_2H_2—air
	239.9	20.0		C_2H_2—air
Cd	228.8	0.03	0.01	C_2H_2—air
	326.1	20.0		C_2H_2—air
Co	240.7	0.16	0.07	C_2H_2—air
	242.5	0.4		C_2H_2—air
Cr	357.9	0.15	0.005	C_2H_2—air
	425.4	0.5		C_2H_2—air
Cs	852.1	0.5	0.05	C_2H_2—air
	455.6	20.0		C_2H_2—air
Cu	324.8	0.1	0.005	C_2H_2—air
	327.4	0.2		C_2H_2—air
Dy	421.2	0.7	0.2	N_2O—C_2H_2
Er	400.8	0.85	0.1	N_2O—C_2H_2
Eu	459.4	0.75	0.2	N_2O—C_2H_2
Fe	248.3	0.14	0.06	C_2H_2—air
	252.3	0.2		C_2H_2—air
	372.0	1.0		C_2H_2—air
Ga	287.4	3.0	0.07	C_2H_2—air
	294.4	4.0		C_2H_2—air
Gd	368.4	17.0	4.0	N_2O—C_2H_2

Table 13.3 — Continued

Element	Line nm	Sensitivity p.p.m.	Limit of detection p.p.m.	Gas mixture
Ge	265.1	2.5	1.0	$N_2O-C_2H_2$
Hf	307.3	15.0	15.0	$N_2O-C_2H_2$
Hg	253.7	10.0	0.2	C_2H_2-air
Ho	416.3	1.4	0.3	$N_2O-C_2H_2$
In	303.9	1.0	0.05	C_2H_2-air
	325.6	1.0		C_2H_2-air
Ir	285.0		4.0	C_2H_2-air
K	766.5	0.03	0.005	C_2H_2-air
	404.4	5.0		C_2H_2-air
La	392.8	42.0	8.0	$N_2O-C_2H_2$
Li	670.8	0.03	0.005	C_2H_2-air
	323.3	15.0		C_2H_2-air
Lu	331.2		50.0	C_2H_2-air
Mg	285.2	0.015	0.000 5	C_2H_2-air
	279.5	5.0		C_2H_2-air
Mn	279.5	0.05	0.005	C_2H_2-air
	280.1	0.15		C_2H_2-air
	403.1	0.08		C_2H_2-air
Mo	313.3	1.5	0.1	C_2H_2-air
	317.0	1.7		C_2H_2-air
Na	589.0	0.03	0.005	C_2H_2-air
	332.0	5.0		C_2H_2-air
Nb	405.9	40.0	20.0	$N_2O-C_2H_2$
Nd	463.4	10.0	2.0	$N_2O-C_2H_2$
Ni	232.0	0.2	0.05	C_2H_2-air
	231.0	0.4		C_2H_2-air
	341.5	3.0		C_2H_2-air
Pb	217.0	0.5	0.01	C_2H_2-air
	283.3	0.7		C_2H_2-air
Pd	247.6	0.8	0.5	C_2H_2-air
Pr	495.1	13.0	10.0	$N_2O-C_2H_2$
Pt	265.9	2.0	0.5	C_2H_2-air
Rb	780.0	0.2	0.005	C_2H_2-air
	420.2	10.0		C_2H_2-air

Table 13.3 – Continued

Element	Line nm	Sensitivity p.p.m.	Limit of detection p.p.m.	Gas mixture
Re	346.0	20.0	1.5	$N_2O-C_2H_2$
		25.0		$O_2-C_2H_2$
Rh	343.5	0.3	0.03	C_2H_2–air
Ru	349.9	2.0	0.3	C_2H_2 air
Sb	217.6	0.5	0.2	C_2H_2–air
	231.1	1.5		C_2H_2–air
Sc	391.2	0.5		C_2H_2–air
Si	251.6	1.2	0.1	$N_2O-C_2H_2$
Se	196.1	3.0	1.0	C_2H_2–air
	204.0	5.0		C_2H_2–air
Sm	429.7	8.5	5.0	$N_2O-C_2H_2$
Sn	224.6	0.7	0.1	H_2-O_2
		2.1	0.4	C_2H_2–air
	286.3	1.3	0.2	H_2-O_2
		3.8	0.6	C_2H_2–air
Sr	460.7	0.15	0.01	C_2H_2–air
	407.8	3.5		C_2H_2–air
Ta	271.5	30.0	6.0	$N_2O-C_2H_2$
Tb	432.6	7.5	2.0	$N_2O-C_2H_2$
Te	214.3	1.5	0.3	C_2H_2–air
Ti	364.3	1.4	0.2	$N_2O-C_2H_2$
	365.4	1.6		$N_2O-C_2H_2$
Tl	276.8	0.8	0.2	C_2H_2–air
	237.3 (238.0)	2.0		C_2H_2–air
Tm	409.4		0.1	C_2H_2–air
U	358.5		12.0	C_2H_2–air
	351.5	200.0	30.0	$N_2O-C_2H_2$
V	318.4	1.3	0.02	$N_2O-C_2H_2$
W	400.9	35.0	3.0	$N_2O-C_2H_2$
Y	407.7	1.1	0.3	$N_2O-C_2H_2$
Yb	398.8	0.17	0.04	$N_2O-C_2H_2$
Zn	213.9	0.02	0.002	C_2H_2–air
Zr	360.1	20.0	5.0	$N_2O-C_2H_2$

presence of rather a small amount of the interfering element, e.g. a concentration equal to that of the element analysed. Less frequent interferences are seen more easily from Appendix I, p. 275, where lines and bands are listed in the order of increasing wavelengths, or from the sections dealing with the properties of the spectra of individual elements.

Table 13.3 contains similar data for absorption flame photometry. This table includes the lines most frequently recommended and employed [100], [102], [108]: molecular bands are not used in this method. The table includes those values which we believe to be most trustworthy. Readers are also referred to data and tables published up to now in the literature [24], [162], [230], [296], [317], [706].

Now, let us discuss the properties of the spectra of individual elements and their behaviour in the flame in more detail.

13.2 Spectra of the individual elements

13.2.1 Alkali metals

Alkali metals are distinguished by a very simple line spectrum, the resonance lines of which are mainly located in the visible part of the spectrum. They can be excited in a flame of relatively low temperature. The tendency to form molecules in the flame (e.g. OH compounds) is very small, the molecular spectrum is quite insignificant from a practical point of view. On the other hand, due to the low ionisation energies of these elements, ionisation is very easy, especially with higher flame temperatures. Alkali metals also have a tendency to form compounds difficult to dissociate, as have alkaline earth elements. Due to their properties mentioned, the determination of alkali metals is a typical application of flame photometry, and for this reason by the far the most attention has been given them in the literature.

The sensitivity of the determination of all these elements is very good. They can also be determined by means of the absorption technique, in which case radiation interferences are eliminated, although the sensitivity of the determination is lower.

13.2.1.1 *Sodium*

The doublet in the yellow part of the spectrum, in the vicinity of the wavelength 589 nm is used practically exclusively for the determination of sodium. It is very easily excited, flames of a very low temperature are quite sufficient for this purpose. This characteristic radiation is used for qualitative as well as quantitative analysis. In the presence of calcium and strontium in the solution analysed, radiation interference may occur, which however may be eliminated rather easily by addition of phosphate or aluminium salts. With higher flame temperatures the results of the determination may be distorted by the presence of elements of a low ionisation potential, and in the case of very low temperatures the influence of dissociation equilibrium may be very unfavourable. The other sodium lines have no practical significance. The above wavelength, i.e. 589 nm, has also been suggested for use in absorption flame photometry.

13.2.1.2 *Potassium*

The most intense lines of potassium form a doublet located in the red part of the spectrum. Their wavelengths are 766 and 769 nm. Coincidence may take place in the presence of rather large amounts of rubidium, in which case mainly the spectral background increases. The tendency to form molecules is rather greater than in the case of sodium, and likewise the intensity of emission of this element is more sensitive to the free electron concentration in the flame than in the case of sodium. Both wavelengths mentioned have been utilised in absorption determinations.

13.2.1.3 *Lithium*

Only the line 671 nm is suitable for emission or absorption determinations of lithium. In the emission technique, coincidence with the molecular band of strontium causes interference. Among all alkali metals, lithium has the greatest tendency to form molecular compounds – hydroxides. Although this phenomenon cannot be utilised for analytical purposes, lithium forms a transition to the group of alkaline earth elements, to which it is rather similar in this respect. The intensity of the radiation emitted, compared with other alkali metals, is little influenced by variations of the electron concentration in the flame, although the influence of some anions, e.g. phosphates, is manifested to a large extent. On several occassions lithium has been proposed for use as internal standard for the determination of sodium or potassium. We believe that with respect to the rather different behaviour of this element the above proposal is not quite justified. It is justified in part only in those cases where the elimination of the influence of physical properties of the solution is involved, specially that of the solvent, while lithium is less suitable in those cases where the removal of the influence of the free electron concentration or of the flame temperature is involved.

13.2.1.4 *Rubidium*

Practically, rubidium can be determined with the use of the lines 780 and 794 nm only. The presence of potassium in the sample may cause difficulties. Coincidence takes place, especially in the short-wave component of the doublet, but with greater potassium concentrations the longer-wave line is affected likewise. It is therefore recommended to use the line of longer wavelength for analytical purposes, although its intensity is less than that of the shorter-wave component of the doublet. The intensity of the two rubidium lines is less sensitive to variations in the concentration of free electrons in the flame and to the flame temperature. Under normal circumstances, rubidium is ionised to a large extent, and therefore the addition of another easily ionised alkali metal has been used with success to increase the emission intensity. In other respects rubidium is similar to potassium. The lines mentioned have also been suggested and used for absorption flame photometry. The advantage is that in the absorption determination radiation interference of the potassium doublet plays no role.

13.2.1.5 *Caesium*

The most intensive caesium line is a doublet of wavelengths 850 and 895 nm, of which only the shorter wavelength component is practically significant. The two lines are practically free of coincidence, it is only essential to take care to avoid radiation interferences, which might take place for example in the presence of excess Ba, Ca, sometimes also Na, etc. The properties of caesium are similar to those of potassium and rubidium, the degree of ionisation is likewise high. Suppression of ionisation contributes to a great extent to increasing the emission intensity of this line. To determine caesium, flames of very high temperatures are not very suitable. The best line for the absorption determination is 850 nm, as most authors state, while the line 455.6 nm is less sensitive.

13.2.2 Alkaline earth elements

These are a very important group of elements, and their determination by flame photometric means receives as much attention in literature as the determination of alkali metals. Most important are the emission determinations of calcium, strontium, and barium. According to the periodic system, the group of alkaline earth elements also includes magnesium, but this element resembles the other elements mentioned only slightly in terms of behaviour in the flame. While the spectra of alkaline earths are among the most intense ones, and they can be determined easily with the use of a flame photometer, magnesium not only forms molecules which dissociate with great difficulty (in this respect, magnesium is similar to the other alkaline earth elements), but its atomic and molecular radiation is very weak due to the influence of high excitation energies in flames of lower temperatures, and this radiation is practically unusable for analytical purposes. Therefore, the following discussion of alkaline earth elements does not apply to magnesium. Its behaviour in the flame and the spectrum it emits will be discussed in a separate section at the end of this chapter.

In flame emitted spectra of alkaline earths lines corresponding to neutral atoms may be identified, as well as complicated systems of bands emitted by molecules containing these elements. The wavelengths of all the lines of calcium, strontium and barium are known with sufficient accuracy and the respective data, including their excitation energy values, are recorded in spectral atlases.

On the other hand, data relating to the character and origin of bands are less satisfactory. Most authors [510], [564] ascribed these bands to molecules of alkaline earth oxides, as it has been proved that roughly 90 % of all calcium atoms, and similarly with strontium and barium, are present in the flame in the form of oxides [389], which are in equilibrium with neutral atoms. Band spectra have also been ascribed to the molecules Ca_2, Sr_2, and Ba_2 [438], [490], or to CaCl [564]. The existence of CaCl in the flame is improbable, since – as shown by HULDT and LAGERQUIST [389] – this molecule dissociates completely. It was shown recently [465] that hydrogen is present in molecules which emit molecular spectra in the visible region: these bands may, therefore, be ascribed to molecules of $Me^{II}OH$, i.e. CaOH, SrOH, and BaOH, the lifetime of which probably is very short, although it suffices for the emission of very intense radiation. According to JAMES and SUGDEN [410] less intense bands are at the same time formed by alkaline earth oxides: these, however, are located in a region of shorter wavelengths.

Although this view is now held generally, it cannot alter the fact that various forms of equilibrium may exist between the different possible forms of alkaline earths in the flame: therefore, alkaline earth spectra are very complicated, and depend to a large degree on the respective experimental conditions. Probably this is also the reason why data on the wavelength of intense peaks of different bands differ somewhat, although these differences are not large. When instruments with filters are employed, precise data on the band edges are not very important, since rather wide wavelength regions are always measured. Disagreement between data given in studies of a practical nature may be ascribed to individual properties of instruments or to slightly differing maximum transmittance values of individual filters.

The possibility of determining these elements by means of absorption flame photometry has also been studied, the result being roughly the same as with emission flame photometry. It must be mentioned in connection with this that the difficulties related to the complicated spectra of alkaline earth elements, and their relatively great tendency to form compounds dissociating with difficulty, are similarly unfavourable in the absorption technique, so that no advantage is gained in this respect.

13.2.2.1 *Calcium*

In the determination of calcium, measurements are usually carried out at three different wavelengths. Firstly, 422.7 nm, i.e. the calcium resonance line; secondly, the maximum of a complicated band at 554 nm; and thirdly, the region used most often with filter-type instruments, 622 nm, where an intensive molecular band is located. Literature reports on the question of which of these bands is most advantageous and allows the most sensitive determination to differ widely. For example, SCHUHKNECHT and SCHINKEL [686] have found the ratio of 100 : 10 : 2 for the emission intensity of calcium at the wavelengths 622 nm, 554 nm and 422.7 nm, respectively. CHEN and TORIBARA [133] believe the radiation corresponding to the wavelength 554 nm to be the most intense. Authors working with monochromator instruments usually prefer the line 422.7 nm, and BURRIEL—MARTI and RAMIREZ—MUNOZ [113] believe this line to be more sensitive than the band spectrum. A comparison of this kind, however, is difficult. An instrument with filters usually records the radiation in rather a broad spectral region, so that the band spectrum will appear to be the most intense one. When a narrow wavelength region only is being measured, a line may be found to be more sensitive. Even in this case, however, much depends on the slit width employed.

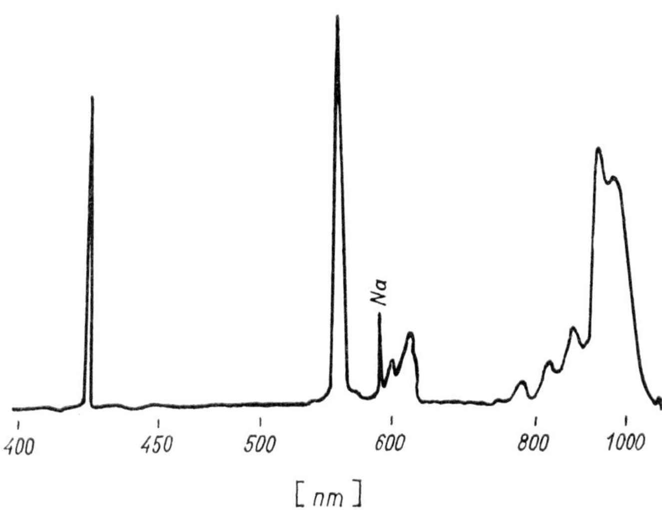

Fig. 13.1. Spectrum of calcium

This complicated nature of the spectrum also has its advantages, since coincidence or other interference in one region may be offset by successful measurement at different wavelengths. Coincidence is an important criterion for selecting the wavelength to be employed. When calcium is determined at 622 nm, an excess of strontium and sodium interferes. Large amounts of sodium make the determination practically impossible. The edge of the 554 nm band cannot be used with high barium concentration for the same reason. According to data published by some authors, the presence of potassium in the sample also contributes to the background. The 422.7 nm line is not directly influenced by any of the alkaline earth elements: it is true that some chromium lines are located in the immediate vicinity, but these are not too intense and they are easily resolved by a good monochromator. If the burner is incorrectly adjusted, the molecular spectrum of some groups will be found in this region, e.g. CH or C_2, at elevated temperatures CN: all these are products of the combustion of gases. From the coincidence point of view, therefore, it is better to use the line than the bands, while the analytical utilisation of calcium bands is hindered to some extent by the circumstance that many elements form a very strong background in the region in which the peaks of these bands are located.

Data concerning the influence of other elements will also be found in the literature, although unfortunately these cannot be arranged on the basis of a unified viewpoint. With higher flame temperatures the spectrum also contains the lines of elements which are difficult to excite, and these may either directly coincide or contribute to the background: at lower temperatures this influence need not be felt at all. These data, therefore, must be accepted with some reservation; the possible influence of differing experimental conditions must be considered, and practical applications must be carried out in accordance with one's own experience.

The line Ca 422.7 nm is always used for absorption determinations: some authors also mention a less sensitive line at 239.9 nm.

13.2.2.2 *Strontium*

The only usable strontium line is located at 460.7 nm. Intensity maxima of the molecular band are found at 666 nm, 680 nm and 605 nm. As with calcium, there is little agreement in views on the relative intensities at these wavelengths. Schuhknecht and Schinkel [686] state an intensity ratio of 4 : 1 for the wavelengths 670 and 460.7 nm, respectively: other authors [136], [182] prefer to measure at 460.7 nm; Gilbert [312] mentions the wavelength 605 nm as being the most intense one. Measurement at a wavelenth of 670 nm may be influenced to some degree by the molecular spectrum of calcium, and the presence of lithium in the sample also makes it impossible to use the wavelength of 760 nm.

Molecular radiation with a peak at 605 nm is in the close vicinity of the sodium doublet and coincides with the intense calcium band. The line 460.7 nm is free of coincidence, but a number of elements increase the background in this region.

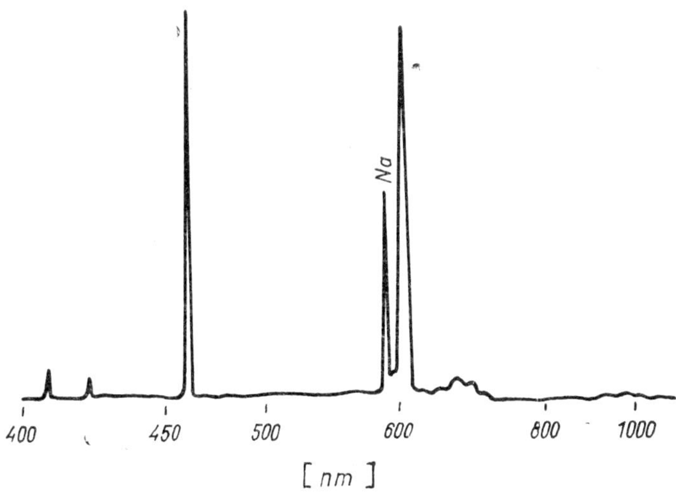

Fig. 13.2. Spectrum of strontium

For absorption flame photometry the line at 460.7 nm is used most often: the ionic line 407.8 nm is less suitable.

13.2.2.3 *Barium*

Of the three typical alkaline earth elements, barium is most difficult to determine, because its radiation is less intense than that of the other two elements, while it is most subject to interference by coincidence and increase of the spectral background. The line 553.6 nm was suggested for analytical purposes, but unfortunately it coincides with an intense calcium band. If the edge of the band in the red part of the spectrum at 840 nm is used, the determination may be complicated by a number of elements which increase the background in this spectral region. Firstly these include the alkalis – sodium and potassium. Potassium also has an adverse influence on measurement at a wavelength of 515 nm, and in connection with this wavelength some authors mention a possible similar interference by magnesium. In the absorption technique, the line at 553.6 nm is used, the other lines are not sufficiently sensitive.

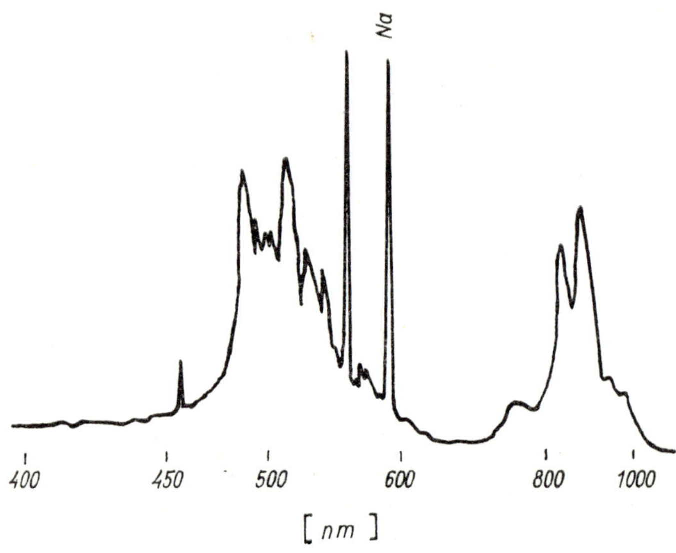

Fig. 13.3. Spectrum of barium

13.2.2.4 *Magnesium*

The emission determination of magnesium is far more difficult than the emission determination of alkalis or the typical alkaline earth elements, mainly because the most intense line of magnesium at 285.2 nm, as well as the band edges at 371 and 384 nm lie in the ultraviolet spectral region. For their determination, a monochromator with quartz optics or a grating is required, only the band with a peak at 384 nm may sometimes be determined with the use of a filter, though obviously with a low sensitivity. The line at 285.2 nm is a resonance line, but it is difficult to excite: a flame of rather a high temperature must be used for this purpose. Equally difficult to excite is the molecular spectrum corresponding to MgOH molecules. Higher temperatures are accompanied by a substantial increase of the background. The line at 285.2 nm is fairly free from coincidences. On the other hand, the two bands named coincide with many elements, e.g. Fe, Tl, etc.

Besides the relatively high excitation energy required, the analysis of Mg is also made difficult by the circumstance that Mg forms a number of thermally very stable compounds which dissociate or volatilise with difficulty. This tendency is pronounced in the presence of certain solution components, e.g. phosphate or aluminium. For this reason, some authors recommend Mg be determined only after separation from the accompanying elements: however, this procedure is not very useful as a rule.

Some possibility of determination is offered by the molecular spectrum of HF, which forms in the F_2-H_2 flame [144] and is distinguished by a peak at 359.4 nm, although it also has distinct peaks at shorter wavelengths. In the hydrogen-peroxo-chlorofluoride flame [676], [693] we find these bands accompanied by the very intense spectrum of MgCl with a peak at 377 nm. In both cases the MgOH spectrum is either completely suppressed or it is very weak. The bands mentioned are more intense than the MgOH bands in other types of flames, although their analytical application is difficult and has not yet been put into practice.

The determination of magnesium by means of absorption flame photometry was more successful. The line used was 285.2 nm. In this case the difficulties related to excitation and coincidence are avoided. Even with this procedure, however, the difficulties which are related to imperfect dissociation of Mg compounds are not negligible.

13.2.3 Copper, silver, gold

All these elements have very simple spectra, with the most intense lines lying in the ultraviolet region. Their tendency to form molecular compounds in the flame is relatively slight, and they do not form any thermally stable compounds, so that there is rather little interference with their determination (with the exception of radiation interference from other elements). The excitation energy of the most intense lines, however, is relatively very high, so that these lines are not very sensitive in the emission method of analysis: better results are obtained by means of the absorption technique.

13.2.3.1 *Copper*

In the copper spectrum, only two lines deserve to be mentioned — 324.8 nm and 327.4 nm, both of which are resonance lines and belong to the same doublet. In the flame, copper is dissociated into atoms to a large extent. In the molecular spectrum, the bands caused by the radiation of CuOH in the green and orange regions of the spectrum are important.

The two lines are adversely influenced by the molecular spectrum of OH, which increases the spectral background: the line of longest wavelength may coincide with the silver line. The use of absorption flame photometry may lead to somewhat better sensitivity: here, the lines at 324.8, 327.4, and 217.9 nm are used most often.

13.2.3.2 *Silver*

With silver, only two lines are of analytical importance, i.e. 328.0 and 338.3 nm, both of which are roughly equally intense. The molecular spectrum is very little developed. The line of shorter wavelength may coincide with a copper line, or it may be subject to interference by the molecular spectrum of OH, while the second line is virtually free from coincidence. The sensitivity of the determination is not very great. The use of absorption flame photometry is far more advantageous: in the literature it is stated that the most sensitive line is the one at 328.1 nm, while the line 338.3 nm is rather less sensitive.

13.2.3.3 *Gold*

Gold has only one line that may be considered useful in emission flame photometry, i.e. the line 267.6 nm. However, nothing has yet been reported about the practical application of this line. The absorption technique is far more advantageous: the line 242.8 nm appears to be more sensitive in this case.

13.2.4 Zinc, cadmium, mercury

These elements are likewise difficult to excite, and they have not yet been determined practically by means of flame photometry, although their spectra and possible applications have been studied. One of the difficulties involved is the fact that the radiation of their resonance lines lies at the limits of transmittance of quartz optics. Good experience has been gained by the use of atomic absorption. Since the elements mentioned do not form compounds in the flame, there is practically no interference in their determination.

13.2.4.1 *Zinc*

There are hardly any practical possibilities of determining zinc by means of emission methods, as the excitation energy of the resonance line 213.8 nm is too great. The line was used, however, with good success and sensitivity in absorption flame photometry. The line at 307.8 nm is far less sensitive.

13.2.4.2 *Cadmium*

The resonance line at 228.8 nm is unusable because of its great tendency to self-reversal, and similarly the intense line at 326.1 nm is not sufficiently sensitive in the emission method: moreover, it may coincide with the radiation of copper, and the background may be increased by the influence of OH bands. The situation is better in the case of the absorption technique, where the line 228.8 nm is the most useful one: the line 326.1 nm is inferior.

13.2.4.3. *Mercury*

The possibility of determining mercury by means of emission or absorption methods has been rather problematic up to now, as none of the methods is sensitive enough. Absorption flame photometry is somewhat more satisfactory, although the sensitivity achieved is inferior to that obtained with other elements. The line at 253.7 nm is mentioned in the literature, as the resonance line Hg 184.9 nm is already located in the region of atmospheric absorption.

13.2.5 Scandium, yttrium, lanthanum and rare earth elements

Up to now we have been unable to determine these elements in a satisfactory way by either emission or absorption methods. Due to their slight dissociation in the flame the emission spectrum of these elements is rather complicated, being mostly composed of bands. These elements are quite easy to identify when isolated, and it might also be possible to consider their emission flame-photometric determination with good sensitivity. The difficulty is that they usually occur in low concentrations and in mixtures. Isolation of the individual elements by chemical means for analytical purposes is rather difficult: coincidences occur in the mixtures to such an extent that it is difficult to determine one element in the presence of another. The most important lines in the bands of these elements are given in Table 14.7. The reader will find information about the possibilities of their analytical application in the original literature [523], [562], [572], [573], [586], [622].

Absorption methods have not yet been applied practically, as they require special conditions [253]. Recently the spectra of the individual elements were studied in rather great detail with the use of dinitrogen oxide and acetylene as fuel mixture: it was found that in this way quite a number of elements can be determined with good sensitivity [497]. However, atomic absorption has not yet been practically applied to the determination of these elements.

13.2.6 Boron, aluminium

Both elements dissociate in the flame with difficulty, and therefore they are not really suitable for absorption flame photometry. While the emission molecular spectrum of aluminium compounds also is very indistinct, boron forms compounds in the flame, the radiation of which may be used for quantitative determinations.

13.2.6.1 *Boron*

In the flame, boron is mainly present in molecular form. It emits a band spectrum with an intensity maximum in the green part of the spectrum: rather important band edges are located at the wavelengths 492, 518 and 546 nm. A mixture of water and methanol is very advantageous for emission, the intensity being increased up to twenty times according to literature data. Barium, as well as chromium, cobalt, iron, and nickel, may interfere with the radiation mentioned. It is an interesting fact that according to literature data, sulphuric acid contributes substantially to increasing the intensity of the molecular radiation of boron [177]. For atomic absorption the line at 249.7 nm has been suggested.

13.2.6.2 *Aluminium*

As already mentioned in respect of attempts at determining aluminium, its flame photometric determination has not been very successful up to the present. Although lines and bands emitted by this element or its compounds are known, they are insufficiently sensitive for emission or for absorption analysis. For the sake of completeness, let us mention that the most intense line is the one at 396.2 nm while the bands have distinct peaks at 467, 484 and 516 nm. With respect to the great chemical stability of the AlO molecule in the flame, a high temperature obviously is required for this spectrum to be obtained in a flame, i.e. at least an acetylene-oxygen flame, or a solution of aluminium in a suitable organic solvent may be used for the analysis [204]. Calcium and iron may coincide with the band at 484 nm. The line at 309.3 nm has the best absorption properties: the dinitrogen oxide – acetylene flame is recommended for the analysis.

13.2.7 Gallium, indium, thallium

The properties of these elements are rather favourable for their flame-photometric determination, as shown by a number of studies. Since their compounds do not dissociate completely, so that molecular forms of these elements are present in the flame, and the excitation energy of their resonance lines is rather high, hotter flames are more suitable for the determination of these elements. There are very few reports in the literature concerning the absorption-photometric determination of these elements, although due to a certain similarity with aluminium and the low degree of dissociation of the elements mentioned, it appears that the sensitivity of the absorption analysis will be only little better than that of the emission method.

The flame spectra of all three elements are very simple, and there are not many lines suitable for analytical purposes. All of them are atomic lines in the blue and long-wave ultraviolet parts of the spectrum. All these elements, especially Ga and In have a great tendency to form hydroxides in the flame, but there is no system of bands suitable for analytical application.

13.2.7.1 *Gallium*

The flame spectrum of gallium includes only two usable lines at wavelengths of 417.2 nm and 403.3 nm. The long-wave line is rather more intense than that of shorter wavelength. The intensity of these lines varies, depending on the part of the flame in which the measurement is being carried out. BULEWICZ and SUGDEN [106] explain this by the fact that a GaOH molecule is formed and exists in the flame. The line at 417.2 nm is subject to no direct coincidence; the line at 403.3 nm may coincide with potassium and manganese. The line at 280.4 nm is usually stated to be most suitable for atomic absorption.

13.2.7.2 *Indium*

The spectrum emitted in a flame by indium includes firstly two lines in the blue part of the spectrum, at wavelengths of 451.1 and 410.2 nm. Again, the long-wave component is rather more intense. Similarly to gallium, the presence of an OH compound in the flame is highly probable, and a weak molecular spectrum between 423 and 430 nm is ascribed to it: however, this is of no analytical significance. The lines mentioned are practically free of coincidence: some authors mention an increase of the background in the presence of excess copper in the solution analysed. For absorption flame photometry, the line at 304 nm was suggested as being the most sensitive one.

13.2.7.3 *Thallium*

Thallium has only two lines which can be used in flame photometry: the first is found at 377.6 nm, the second at 535.0 nm. The first of the two is rather more sensitive. Its disadvantage, however, is the fact that it can easily coincide with the molecular radiation of MgOH. The tendency to form compounds in the flame is substantially less in the case of thallium, compared to gallium and indium: no spectrum corresponding to an OH compound was found, although some authors mention the possibility of formation of a weak spectrum of TlH in the yellow and orange spectral regions. From the analytical point of view this is quite unimportant. Thallium appears to be more sensitive when analysed by absorption techniques. The lines at 276.8 and 377.6 nm are mentioned as being suitable.

13.2.8 Group IV Elements

These elements are not often determined by flame photometric means, especially by the emission method. There is some possibility of utilising chemiluminescence. In absorption analysis, some success was achieved with the determination of lead, while the methods proposed up to now for the determination of tin and titanium are less sensitive.

13.2.8.1 *Lead*

The wavelengths of the most intense lines of lead are 368.4 and 405.8 nm: weak molecular bands of PbO were also found in the spectrum, although up to now they have not been utilised for analytical purposes. The lead line 405.8 nm may coincide with lines of manganese, potassium, possibly also gallium. Lines of wavelength 217.0 nm, in some cases also 283.3 nm were used for absorption flame photometry.

13.2.8.2 *Tin*

Tin has no lines which could be excited in a simple flame to such a degree that they could be employed for analytical purposes. The molecular spectrum of tin in the flame is unfortunately weak to such an extent that it cannot be put to any practical use. On the other hand, a flame with an excess of acetylene gives relatively intense radiation at wavelengths of 270.6 and 242.9 nm. The lines 224.6 and 286.3 nm are recommended for absorption flame photometry, though the sensitivity of the determination is substantially less than with other elements. The sensitivity may be improved by adjusting reducing conditions in the flame [296].

13.2.8.3 *Titanium*

Titanium is not very suitable for either emission or absorption analysis. In the flame it forms a molecular spectrum with no distinct peaks, rather similar to a continuum. Some possibility of determination is offered by absorption flame spectrophotometry in a hot flame – dinitrogen oxide + acetylene. The wavelength 364.3 nm is stated to be most suitable for measurement.

13.2.9 Group V Elements

The emission spectrum of these elements in the flame is rather indistinct, so that it has found no practical application. Somewhat better were the results of experiments with absorption measurement, where quite a high sensitivity was achieved, obviously with the use of hot flames.

13.2.9.1 *Vanadium*

The line spectrum of vanadium is not very distinct either in the hydrogen or in the acetylene flame. A higher temperature is required to obtain a spectrum which could be used for analysis. At lower temperatures, bands with peaks at 523, 547 and 574 nm are formed, but the sensitivity is not too good. The situation is similarly disadvantageous in absorption flame photometry, where the line at 389 nm as well as other lines were tested, although the results have not been good up to now and inferior sensitivity values were recorded. There is some hope in the use of a flame with an excess of acetylene, in which a relatively sensitive line was found at 318.4 nm [703].

13.2.9.2 *Niobium*

The line at 405.9 nm is stated to be useful for emission as well as absorption measurement. Emission, which is more distinct in this element than absorption, takes places in the inner zone of the flame.

13.2.9.3 *Tantalum*

Among the many lines of the tantalum spectrum, the one at 481.3 nm with rather low sensitivity is usually recommended: the line at 271.5 nm is only slightly better for absorption.

13.2.9.4 *Arsenic*

This element is not suitable for emission analysis at all, since its spectrum is characterised by very weak and practically unusable lines. Atomic absorption with the use of the line 193.7 nm is rather promising.

13.2.9.5 *Antimony*

Among the very weak antimony lines, the one at a wavelength of 259.8 nm appears to be most suitable for emission. This element may, however, be determined with better sensitivity by means of atomic absorption, in which case the line at 217.6 nm should be used. Another somewhat inferior line is located at 231.1 nm.

13.2.9.6 *Bismuth*

As with the preceding elements, atomic absorption should be considered as a practical possibility for analysis, as the emission spectrum is very weak. The most suitable line is stated to be 223.1 nm, although according to some literature data the lines 222.9 and 306.8 nm are almost as useful.

13.2.10 Selenium, tellurium, tungsten

At present these elements cannot be determined by means of emission flame photometry, as even under specially selected donditions they do not form a spectrum of a character which would make it suitable for analytical purposes. It seems that atomic absorption is more promising. The sensitivities achieved are not good, although in theory the elements can be determined. To this purpose, a line at 196.1 nm or 204.0 nm is recommended for selenium, for tellurium at 214.3 nm, and lastly, for tungsten at 400.9 nm.

13.2.11 Chromium, molybdenum, manganese

These elements are characterised by atomic spectra with many lines, which are not very intensive but, with a suitable flame temperature, permit the elements named to be determined by means of emission flame photometric methods with a sensitivity far superior to that of the elements of the preceding group. An exception is molybdenum.

A disadvantage of these elements is their rather great tendency to form relatively stable oxygen-containing compounds in the flame, which make it impossible to utilise the relatively low excitation energy of many lines. Substantially better sensitivity is obtained with absorption methods.

13.2.11.1 *Chromium*

The most intense lines of chromium lie in the blue part of the spectrum: the most important one is the line 425.4 nm. The line in the ultraviolet spectral region (357.9 nm) is far weaker. The line at 520.8 nm also has some analytical possibilities. The molecular spectrum of chromium is of no value in emission flame photometry. The line at 425.4 nm may coincide with iron, an intense calcium line is located very close to it.

The lines in the ultraviolet part of the spectrum may be influenced unfavourably by the molecular spectrum of CN. The line 357.9 nm was suggested as being the most suitable one for absorption measurement: the determination with the use of the line 425.4 nm is less sensitive.

13.2.11.2 *Molybdenum*

Molybdenum trioxide dissociates with very great difficulty, so that there are no distinct lines of molybdenum which could be used for analysis. We may rather speak of a continuum exhibiting relatively indistinct peaks at wavelengths close to 532 nm. The situation is somewhat better when a cyanogen-oxygen flame is used: the most sensitive line is found at 390.3 nm, but it is influenced unfavourably by the CN band. The line at 313.3 nm was proposed for absorption, as well as the line 317.0 nm, which however is less sensitive.

13.2.11.3 *Manganese*

Manganese is somewhat easier to determine than the elements discussed above. It is characterised by rather an intense triplet at 403 nm, which unfortunately may coincide with the radiation of potassium and gallium. The triplet at 279 nm is a little less intense. Manganese also emits a relatively complicated molecular spectrum. It has peaks at wavelengths of 559 and 539 nm. However, in neither of these cases is sensitivity as good as in the case of the absorption determination at wavelengths of 279.5, 280.1 or 403 nm.

13.2.12 Iron, cobalt, nickel

These elements are also characterised by a spectrum composed of numerous lines, the most intense of which are in the ultraviolet part of the spectrum. They have a relatively great tendency to form compounds in the flame, though their molecular spectra however are not very distinct. The relatively low degree of dissociation at lower flame temperatures is a disadvantage with emission as well as absorption methods. Nonetheless, the absorption method is a little more sensitive.

13.2.12.1 *Iron*

It is well known that the spectrum of iron is unusually rich in lines, although lines of sufficient sensitivity are only found at high flame temperatures. These mainly include the wavelengths 372 and 386 nm. The first of these may coincide with manganese and thallium, while the line of greater wavelength may also coincide with the molecular spectrum of magnesium. In the visible spectral region we also find bands which probably correspond to FeO molecules: these, however, are more important because they may have an adverse effect on the determination of other elements by increasing the background radiation, if iron is present in a large concentration. In absorption, the most sensitive lines are 248.3 nm and 252.3 nm.

13.2.12.2 *Cobalt*

Cobalt has similar properties to iron. The lines at 350, 352 and 345 nm are the most intense ones in emission. They are relatively free of coincidence, although the OH spectrum must be taken into account, as it may also penetrate into this region. The cobalt lines are very often close to each other to such an extent that they are difficult to resolve by means of conventional spectrophotometers. The lines at 240.7, 242.5 and 345.3 nm have been proposed for absorption measurement.

13.2.12.3 *Nickel*

Among the numerous lines of nickel, the ones at 352, 341 and 351 nm are most intense. Flames of a high temperature must be used for their excitation. The molecular spectrum is weak in the ultraviolet as well as in the blue and green parts of the spectrum. There is a possibility of coincidence with the lines of cobalt and iron. The lines at 232, 231 and at 341 nm are used chiefly for absorption.

13.2.13 Platinum-group metals

These elements form a relatively intense emission spectrum of the line or band type at elevated temperatures, but only little practical application of these spectra has been made up to now. For the sake of completeness let us mention the lines Pd 340 and 363 nm, Ru 372 and 379 nm, Rh 343 and 369 nm and the band peak at 542 nm formed by the RhO molecule. Absorption analysis, however, is more practicable: suitable lines are Pd 244, 247 nm, Rh 343.5 and 369 nm, Pt 265 and 217 nm. The absorption method has the advantage of perfect selectivity.

13.2.14 Non-metals

Non-metals in general are known to be difficult to excite, even in energy sources of a temperature substantially superior to that of flames. The atomic spectrum of these elements therefore is not emitted in flames. However, a molecular spectrum may be obtained by rather conventional means with the majority of non-metals, if a suitable element is chosen to form a compound with the element to be analysed. For example, the very intense spectrum of $CuCl_2$ is known to have peaks at the wavelengths 435 and 473 nm [502]: this spectrum was proposed for the determination of chlorine. For the same purpose, the molecular spectrum of InCl with a peak at 359.6 nm is even more suitable, as stated by GILBERT [319]. Similarly, the intensity of CN bands is proportional, to a large extent, to the nitrogen concentration in the sample (obviously if suitable experimental conditions are chosen). Bands of OH or CO, as well as of C_2, are very intense. At present, a general procedure for the determination of these elements cannot be suggested, but it is probable that in special cases the properties of these molecular spectra might be utilised for analytical purposes. No attempts at a direct determination of sulphur and phosphorus have been made, although several procedures are known for indirect determination of these elements: however, these procedures are not based on the properties of the spectra of the elements to be determined. At present, these elements are thought to be unsuitable for absorption flame photometry.

14. Practical applications of flame photometry

The applicability of flame photometry, like that of other physico-chemical methods, depends to a large extent on the nature of the sample in which the individual elements are to be determined. If we were to take into account the fact that the composition of every sample will be different, we should—strictly speaking—have to work out a special procedure for every sample, and there would be no sense in giving analytical procedures in this book.

With the contemporary state of development of industry, with typification and standardisation, however, many products, raw materials, etc., which are important from a practical point of view, have rather a constant composition. This permits a wide field of application of flame photometry in routine control analyses of industrial products, raw materials, intermediates, etc. In such cases, published procedures, which have been worked out for some specific type of material, may be utilised. Sometimes, however, the composition of the material to be analysed differs widely from the material for which the procedure has been worked out. In this case the procedure generally has to be modified. Of course, each alteration of the procedure should be tested beforehand.

Thus, those procedures which describe the determination of a specific component in a specified technical material are of particular practical significance. Therefore we are presenting procedures for the determination of individual elements in the most varied materials in this chapter. From these procedures, the difference between them for the determination of an element in various materials will also follow. For obvious reasons, we have only been able to include a few examples of the application of flame photometry in this book. We have therefore selected mainly those procedures which in our opinion are capable of being applied most widely. We are mentioning other possibilities by means of referring to the literature. We have tested the majority of the methods described practically, and are utilising the experience thus gained in these procedures. This also explains some deviations from the data given in the original literature.

As far as possible we are describing methods for which simple filter-type instruments suffice. In some cases, however, the use of instruments fitted with monochromators cannot be avoided. Frequently also, procedures which have been worked out for one type of instrument are unsatisfactory when a different type is used. Therefore we are also stating the type of instrument with every procedure.

14.1 **Water analysis**

Water samples may have the most varied origin and composition. In some cases (effluents) water may contain large amounts of substances which do not occur in water conventionally and which might influence the analysis (e.g. large amounts of organic substances may influence the emission of many elements, particularly the presence of surfactants, see chapter 12). The analytical procedure must be adjusted to such special conditions. For example, the sample must be suitably prepared for analysis by mineralisation of organic substances, etc. Similarly, the analysis proper must be preceded by filtration or sedimentation to separate dispersed solid substances. If such water samples were dispersed directly, the dispersion system would be clogged, leading on the one hand to errors in the determination, on the other to a loss of time involved in cleaning the apparatus. In this connection it should be stressed that it may sometimes be wrong to separate solids directly from the original sample without any other precaution being taken. For example, when calcium is being determined in mineral water of acid character, there always is the danger of carbon dioxide escaping, with the result that calcium carbonate precipitates after conversion from the more soluble hydrogen carbonate. In this way, calcium might be separated together with the precipitate, escaping the determination. In this case the sample must be acidified and only then can the precipitate be separated.

Depending on the content of the component to be analysed, it may be useful to adjust the sample concentration by dilution or evaporation of the sample. These operations must obviously be carried out in such a way as not to contaminate the sample with the element to be analysed. This danger exists, for example, when distilled water is used to dilute a sample in which sodium, potassium, etc., are to be determined, as the distilled water always contains traces of these elements extracted from glass. This applies to distilled water obtained in a glass apparatus. Therefore it is best to use water distilled from an apparatus of quartz or, better, water which has been deionised by means of ion exchange resins in a column made of polythene. In every case it is recommended to store distilled water in bottles of polythene.

Likewise when a sample is being concentrated by evaporation it is essential to avoid the possibility of introducing the element to be determined. Again, glass is not suitable as material: vessels of teflon, quartz or platinum are better.

The use of flame photometry in the analysis of water is the subject of several review articles [146], [679], [691] and advances in this field are currently being recorded in the journals Analytical Chemistry and Japan Analyst.

Flame photometric methods are being applied most widely to the determination of sodium, potassium, and calcium, which occur in practically all types of water. We are therefore including detailed procedures for their determination.

14.1.1 Determination of sodium and potassium [687]

The following procedure is suitable for samples of natural water and frequently for effluent samples too. The interfering influence of calcium present in the sample is eliminated by an addition of aluminium nitrate (see section 12.2). The addition of aluminium nitrate in a concentration of 25 mg $Al(NO_3)_3 \cdot 9\,H_2O$/ml permits, as our experience indicates, sodium and potassium to be determined in the presence of up to 200 µg Ca/ml. Iron concentrations of more than 50 µg Fe/ml interfere in the determination of sodium.

Solutions

A solution of aluminium nitrate is prepared by dissolving 500 g $Al(NO_3)_3 \cdot 9\,H_2O$ in demineralised water and making up to one litre. The aluminium nitrate employed must be very pure, containing no Na and K.

The standard sodium chloride solution containing 100 µg Na/ml and the standard potassium chloride solution containing 100 µg K/ml are prepared as stated in Appendix II.

The blank solution, containing 25 mg $Al(NO_3)_3 \,.\, 9\,H_2O$/ml, is prepared by diluting 5 ml aluminium nitrate solution to 100 ml (the composition of this solution is identical with that of the solutions NaI_6 and KI_6, see Table 14.1). The measuring solution containing 20 µg Na/ml, 10 µg K/ml, and 25 mg $Al(NO_3)_3 \cdot 9\,H_2O$ in 1 ml is prepared by diluting 20 ml of the standard sodium solution, 10 ml of the standard potassium solution, and 5 ml of the standard aluminium nitrate solution to 100 ml. The composition of this solution is identical with that of solutions $Na3_1$ and $K3_1$ (see Table 14.1).

When large series of samples are to be analysed, it is advantageous to prepare a sufficient amount of the measuring and zero solutions, since the setting of the instrument must be checked by means of these two solutions after every 3 to 5 measurements.

Table 14.1(a). PREPARATION OF STANDARD SOLUTIONS

Volumes of solutions for constructing calibration curves for Na

Notation of solution	Volume of standard Na solution, ml	Volume of standard K solution, ml	Volume of $Al(NO_3)_3 . 9\,H_2O$ solution, ml	Na content/ 100 ml, mg
Na 1_1	20	0	5	2.0
Na 1_2	16	0	5	1.6
Na 1_3	12	0	5	1.2
Na 1_4	8	0	5	0.8
Na 1_5	4	0	5	0.4
Na 1_6	0	0	5	0
Na 2_1	20	4	5	2.0
Na 2_2	16	4	5	1.6
Na 2_3	12	4	5	1.2
Na 2_4	8	4	5	0.8
Na 2_5	4	4	5	0.4
Na 2_6	0	4	5	0
Na 3_1	20	10	5	2.0
Na 3_2	16	10	5	1.6
Na 3_3	12	10	5	1.2
Na 3_4	8	10	5	0.8
Na 3_5	4	10	5	0.4
Na 3_6	0	10	5	0

Instrument

Photometer with dispersion chamber (filter-type, Zeiss)

Conditions

Flame: acetylene – air
Wavelengths measured: Na 589 nm; K 768 nm (interference filters Na 59, K 77 J). Concentration range: 5 to 20 μg Na/ml; 2 to 10 μg K/ml
Method of evaluation: system of calibration curves (see section 10.7.2)

Construction of the system of calibration curves

Three series of solutions are prepared for sodium and three for po-

Table 14.1(b). Volumes of solutions for constructing calibration curves for K

Notation of solution	Volume of standard K solution ml	Volume of standard Na solution ml	Volume of $Al(NO_3)_3 . 9\,H_2O$ solution ml	K content/100 ml mg
K 1_1	10	0	5	1.0
K 1_2	8	0	5	0.8
K 1_3	6	0	5	0.6
K 1_4	4	0	5	0.4
K 1_5	2	0	5	0.2
K 1_6	0	0	5	0.0
K 2_1	10	10	5	1.0
K 2_2	8	10	5	0.8
K 2_3	6	10	5	0.6
K 2_4	4	10	5	0.4
K 2_5	2	10	5	0.2
K 2_6	0	10	5	0.0
K 3_1	10	20	5	1.0
K 3_2	8	20	5	0.8
K 3_3	6	20	5	0.6
K 3_4	4	20	5	0.4
K 3_5	2	20	5	0.2
K 3_6	0	20	5	0.0

tassium with the use of the standard sodium and potassium chloride and aluminium nitrate solutions. The solutions are prepared as shown in Table 14.1, the volumes stated being measured by pipette into 100 ml volumetric flasks, made up to the mark and stirred. Values obtained by measuring these solutions (similarly as in the procedure) are used to construct a system of calibration curves. In constructing these curves, galvanometer deflections are plotted versus the Na or K concentration expressed in mg/100 ml. For individual solutions these values are also given in Table 14.1. Our experience shows that a single calibration curve suffices for sodium, since the three curves obtained for sodium according to Table 14.1 are virtually identical.

Procedure

Into a 100 ml volumetric flask, an amount of the water sample to be analysed corresponding to 0.5 to 2.0 mg Na and 0.2 to 1.0 mg K (usually 50 ml) is measured by pipette. If the water is alkaline, it is acidified with dilute hydrochloric acid to pH 3.5 ml of the aluminium nitrate solution is then added, the solution is made up to the mark with demineralized water and stirred. The zero point of the instrument is set for the blank solution, using the sodium filter, and the maximum deflection is set for the measuring solution by means of the diaphragm. This is done with the maximum sensitivity of the galvanometer. The maximum deflection set must obviously have the same value as in the construction of the calibration curve. When the instrument has thus been adjusted, the sample solution is measured to obtain the deflection for sodium. The potassium filter is then inserted and the entire procedure (i.e. setting the zero and maximum deflections and measuring the sample deflection corresponding to potassium) is repeated.

The deflection of the measuring instrument corresponding to potassium is first used to read the approximate potassium content from the calibration curve, which was obtained by measuring a series of standard solutions containing no sodium (series K 1 in Table 14.1). When the sodium content has been read, that one of the sodium calibration curves is used from among the curves obtained by measuring the series of standard solutions, the potassium content of which is closest to that determined by the above method. The potassium content is then obtained in a similar manner.

When potassium as well as sodium are to be determined in a number of samples, the procedure involves setting the instrument for the measurement of sodium, measuring the sodium emission in all samples, then setting the instrument for potassium and measuring the potassium emission in all samples.

Calculation:

$$\text{mg/l Na (or K)} = \frac{a \cdot 1000 \cdot k}{b},$$

where a is the amount of sodium or potassium read off the calibration curve in mg/100 ml, b is the volume of water sample pipetted for the analysis, in ml, k is a correction factor for the concentration change due to preliminary preparation of the sample by dilution or evaporation: $k = \frac{v_2}{v_1}$, v_1 being the original volume, v_2 the volume after the concentration adjustment. If no such operation was carried out, $k = 1$; if it has been concentrated, $k < 1$; if it was diluted, $k > 1$.

14.1.2 Determination of calcium [635]

The procedure for the determination of calcium described here is applicable virtually universally for various types of water. The interfering influences of alkali metals, aluminium, sulphates, phosphates, and silicates are eliminated by the addition of a buffer solution (saturated sodium chloride and potassium chloride solution containing 50 mg La/ml). When this buffer solution is used, neither sodium and potassium present in amounts of up to 500 μg/ml, nor aluminium, sulphates, or phosphates in amounts of up to 200 μg/ml interfere in the determination of calcium In most cases a buffer solution of the above composition is fully sufficient. In special cases, e.g. in the presence of a greater amount of aluminium, the buffer solution may be modified by increasing the lanthanum concentration so as to eliminate the interference due to aluminium even at the higher concentration.

Solutions

The buffer solution is prepared by dissolving 42 g $La(NO_3)_3 \cdot 6\,H_2O$ in 100 ml saturated sodium and potassium chloride solution. This saturated solution is prepared by letting 50 g NaCl and 50 g KCl stand in 100 ml demineralised water in a polythene vessel with frequent agitation, after which the non-dissolved salts are separated. A standard calcium chloride solution, containing 100 μg Ca/ml is prepared as stated in Appendix II.

The zero solution is prepared by adding 1 ml buffer solution to 25 ml demineralised water and stirring. The measuring solution is prepared by adding 1 ml buffer solution to 25 ml standard calcium chloride solution and stirring. It is identical with the last solution of the calibration series. In the case of routine analyses it is advantageous to prepare a larger amount of the zero and measuring solutions, as the instrument setting must be checked after every two or three measurements.

Instrument

Photometer with dispersing chamber (filter-type, Zeiss)

Conditions

Flame: acetylene–air
Wavelength measured: Ca 622 nm (filter, Zeiss Ca 63 J)
Concentration range: 10 to 100 μg Ca/ml
Method of evaluation: single calibration curve method

Construction of the calibration curve

Volumes of 0, 1.25, 2.5, 7.5, 15 and 25 ml standard $CaCl_2$ solution (i.e. 0, 0.125, 0.250, 0.750, 1.5 and 2.5 mg Ca) are measured by means of a calibrated pipette into volumetric flasks 25 ml in volume and made up to the mark with demineralised water. 1 ml buffer solution is added to each solution by pipette and the solutions are stirred. Then the solutions are measured in the way described in the procedure. A calibration graph is plotted from the measured values, galvanometer deflections being plotted versus the calcium concentration expressed in mg/25 ml.

Procedure

1 ml buffer solution is added to 25 ml of the water sample to be analysed, containing 0.25 to 2.5 mg Ca/25 ml (if required, the concentration of the sample is first adjusted by dilution or evaporation) and the mixture is stirred. The galvanometer zero and the maximum deflection are set for the zero and measuring solutions, respectively, with the calcium filter inserted. The zero and maximum settings are repeated until both values are constant. The sample is then measured and the respective galvanometer deflection is read. The amount of calcium corresponding to the deflection is then read off the calibration curve.
Calculation:

$$\text{mg/l Ca} = a \cdot 40 \cdot k$$

where a is the calcium amount read off the calibration curve in mg/25 ml, k is the correction factor for the change due to the preliminary concentration adjustment (see determination of sodium and potassium, section 14.1.1).

The reproducibility of the calcium determination, established by means of analyses of practical water samples, is $\pm 1.5 \cdot k$ mg Ca/l. This value is stated for a statistical security of 99 %.

Besides the procedures which we are mentioning here, a number of other methods are described in the literature for the determination of sodium [797], [555], [778], [522], [611], [775], [507], [741], potassium [522], [555], [405], [507], [741], [776], [797], and calcium [797], [778], [770], [777], [742], [612], [139], Methods have also been developed for the determination of lithium [506], [769], [320], [132], strontium [778], [741], [770], in sea water [136], [708] and of some other elements [248].

14.2 The application of flame photometry in the chemical industry

Some methods which are suitable for control purposes in chemical production are described in detail in this chapter. Obviously it is impossible to discuss all possible applications within a limited scope. Therefore, let us describe several methods for the analysis of widely varying materials in order to show the wide range of application of flame photometry in the chemical industry.

With respect to the varied character of the materials involved, no general instructions can be given for the preparation, decomposition, etc., of the samples. Here, practical considerations must be based on the principles of general analytical chemistry. Besides the examples described in detail, we will mention other possibilities in references to the original literature.

14.2.1 Determination of potassium in combined fertilisers

Recently, so-called combined fertilisers are being used increasingly: these contain the three main nutrients, potassium, nitrogen, and phosphorus. The determination of these three components is important for testing the quality of the fertiliser. Obviously, flame photometric means can be used with advantage to determine potassium. The influence of calcium, which is always present in the type of fertiliser mentioned, is eliminated in the procedure by the addition of aluminium nitrate. Sodium influences this determination only if present in a concentration greater than that of potassium. In these types of fertiliser, however, sodium occurs in concentrations substantially lower, so that its influence is not felt. The method described offers results which agree well with the determinations by gravimetric analysis with sodium tetraphenylborate.

Solutions

The aluminium nitrate solution is prepared by dissolving 500 g $Al(NO_3)_3 \cdot 9\ H_2O$ in distilled water and making up to one litre. The standard potassium chloride solution contains 1 mg K/ml (see Appendix II). The blank solution is prepared by diluting 5 ml aluminium nitrate solution and 0.2 ml hydrochloric acid (1 to 1) to 100 ml. The measuring solution is prepared by diluting 5 ml aluminium nitrate, 10 ml standard potassium chloride solution, and 0.2 ml hydrochloric acid (1 to 1) to 100 ml. This solution is identical with the last solution of the calibration series.

Instrument

Photometer with dispersing chamber (filter-type, Zeiss)

Conditions

Flame: acetylene – air
Wavelength measured: K 768.2 nm (filter Zeiss K 77 J)
Range of determination: 2 to 40 % K
Method of evaluation: single calibration curve method

Construction of the calibration curve:

Standard potassium chloride solution volumes of 0, 1, 3, 5, 8, 10 ml (i.e. 0, 1, 3, 5, 8, 10 mg K), and 0.2 ml hydrochloric acid (1 to 1), are measured by pipette into 100 ml volumetric flasks and diluted with distilled water to roughly 80 ml. 5 ml aluminium nitrate solution is then pipetted into each flask, and the solutions are made up to the mark and stirred. The solutions are then measured as stated in the procedure. A calibration curve is constructed from the data measured by plotting the galvanometer deflection versus the potassium concentration in mg/100 ml.

Procedure

Roughly 2.5 to 5 g of the finely ground fertiliser sample (depending on the expected potassium content) is weighed into a 500 ml volumetric flask, 100 ml distilled water and 20 ml hydrochloric acid (1 to 1) are added. The contents of the flask are gently heated until a solution is obtained (about 10 minutes). After cooling, the solution is made up to the mark with water, stirred and filtered by a dry fluted filter into a dry beaker. The first filtrate portions are poured away, and from the following ones, a 5 ml aliquot is pipetted into a 100 ml volumetric flask, diluted with distilled water to roughly 80 ml, 5 ml aluminium nitrate solution is added, and the solution is made up to the mark with distilled water and stirred. The galvanometer zero and maximum deflection are set for the zero and maximum measuring solutions, with the potassium filter inserted, and then the ;sample solution is measured. The amount of potassum corresponding to the measured deflection is then read off the calibration curve.

Calculation:

$$\% \mathrm{K} = \frac{10a}{n} \text{ or, } \% \mathrm{K_2O} = \frac{12.05a}{n},$$

where a is the amount of potassium read off the calibration curve, mg/100 ml, n is the sample weight in g.

The reproducibility of the determination, established by determinations of potassium in practical samples, is $\pm 0.3\%$ abs. for K, or $\pm 0.4\%$ abs. for K_2O. This value is stated for a statistical security of 99 %.

Other procedures for the determination of potassium in fertilisers are described in the literature [34], [72], [84], [92], [129], [150], [267 – 274], [301 – 303], [470], [504], [526], [577], [668], [699].

14.2.2 Determination of P_2O_5 in technical trisodiumphosphate [637]

Technical trisodiumphosphate is used in industry as a water softener. For production control as well as for the determination of the required doses, the P_2O_5 content must be known. A rapid determination of the P_2O_5 content is made possible by the method described in the following, based on the work of DIPPEL, et al. [195]. Differing from the authors named, who use an instrument with direct injection of the solution, and a monochromator, we have modified the method so as to permit the use of a filter-type instrument with mist chamber (Zeiss Model III) which is more conventional in many laboratories. To increase the accuracy of the determination the measurement must be repeated at least three times. A small amount of sulphates (1 to 2 %) usually present in samples does not influence the analysis.

The method gives results which agree well with the gravimetric P_2O_5 determination by the Lorenz method.

Solutions

The standard calcium solution containing 10 mg Ca/ml is prepared by dissolving 12.4865 g calcium carbonate, adding a small amount of concentrated hydrochloric acid, evaporating till dry on a water bath, dissolving the residue in water, and making up to 500 ml. The solution is stored in a polythene bottle.

The standard phosphoric acid solution is prepared by diluting 4.8 ml 85 % phosphoric acid to 1000 ml with water. This solution contains roughly 5 mg P_2O_5/ml. The accurate value is determined by titrating 25 ml of this solution with 0.1 M sodium hydroxide solution using dimethyl yellow as indicator.

The measuring solution is prepared by diluting 10 ml standard calcium solution to 100 ml with water. This solution is identical with the first solution of the series which serves for constructing the calibration graph.

Ion exchanger column

The ion exchanger column can be made from old burettes of 50 ml volume. The upper end of the burette is cut away so as to leave sufficient space for 10 ml of the solution when the column is filled with the ion exchange resin. The column is filled with 10 g air-dried cation exchange resin (Dowex 50). The ion exchanger column rests on a plug of glass wool located in the lower conical part of the burette. After every determination the ion exchanger is regenerated with 50 ml 3N hydrochloric acid and washed with water, until the eluate does not have an acid reaction (testing with pH-papers).

Instrument

Filter-type photometer with dispersion chamber (Zeiss Model III)

Conditions

Flame: acetylene-air
Wavelength measured: Ca 622 nm (Filter Zeiss Ca 63 J)
Concentration range: 10 to 30 % P_2O_5
Method of evaluation: single calibration curve method

Construction of the calibration curve

0, 5, 10, 15, 18 and 20 ml standard phosphoric acid solution are measured by pipette into volumetric flasks 100 ml in volume (i.e. 0, 25, 50, 75, 90 and 100 mg P_2O_5). The volume in all flasks is adjusted with water to roughly 85 ml, and 10 ml standard calcium solution is added by pipette to every flask. The solutions are made up to the mark with water and measured as described in the procedure.

The calibration curve must be constructed anew for every series of samples. The standard solutions are measured twice, i.e. before and after the sample solutions are measured. The mean values thus obtained are plotted in a graph versus the P_2O_5 concentration in mg/100 ml.

Procedure

A sample of 6.25 g is weighed into a 250 ml volumetric flask, dissolved in roughly 100 ml water, and the solution is made up to the mark with water and stirred. A 10 ml aliquot is pipetted onto the ion exchange column and allowed to flow at a rate of roughly 5 ml/min. The eluate is collected in a 100 ml volumetric flask. The ion exchanger is then washed with water until the overall eluate volume has reached about 85 ml. 10 ml standard calcium solution is added by pipette, and the solution is made up to the mark and stirred. While these operations are being carried out, the photoelectric cell is allowed to "settle down" by dispersing a solution, containing 5 ml standard calcium solution in 100 ml, for 15 to 20 minutes, with the calcium filter inserted. The zero point is then set at the maximum galvanometer sensitivity for distilled water, and the maximum deviation is set (best at 250 divisions) by means of the diaphragm for the measuring solution. The two settings are repeated several times until the values are constant. The maximum deviation is read in the moment when, after an initial rapid rise of the galvanometer reading the scale stops clearly for the first time and oscillates for several seconds around a certain value. The sample is measured as quickly as possible, immediately after the maximum deviation has been set. The reading is taken in the same way, i.e. the value around which the scale oscillates for some time in the first stabilisation is read.

The maximum deviation must be set before every measurement, while it suffices for the zero point to be checked each time after 3 or 5 measurements. Each solution is measured three times as described above. The mean values are then used to read the required amount of P_2O_5 in mg/100 ml from the calibration curve.

Calculation:

$$\% \, P_2O_5 = \frac{a}{2.5},$$

where a is the amount of P_2O_5 read from the calibration curve, mg/100 ml.

The reproducibility of the method, determined by P_2O_5 analyses in practical samples, is ± 0.55 abs. %. This value is given for a statistical security of 99 %.

14.2.3 Determination of calcium in phosphoric acid made by the extraction process [637]

In the manufacture of phosphoric acid by the extraction process, one of the important factors is the ease of filtration of the mixture of phosphoric acid and gypsum which is formed. The formation of easily filtered crystals depends on the ratio of calcium to sulphate ions in the mother liquor [715]. From the point of view of control of this process, the determination of calcium therefore is also important. The method to be described is based on studies by BRABSON and WILHIDE [89], although it has been modified for use with the Zeiss Model III flame photometer. The interfering influence of aluminium, which is separated together with calcium, is eliminated by the use of a buffer containing sodium, potassium, and lanthanum.

Solutions and Reagents

The buffer solution is made by dissolving 42 g $La(NO_3)_3 \cdot 6\,H_2O$ in 100 ml saturated sodium and potassium chloride solution. This saturated solution is prepared by allowing 50 g NaCl and 50 g KCl to stand for several hours in 100 ml demineralised water in a polyethylene vessel with frequent agitation, and then filtering to remove the non-dissolved salts.

The standard calcium chloride solution of 1 mg Ca/ml is prepared as indicated in Appendix II.

The blank solution is prepared by diluting 5 drops concentrated hydrochloric acid, and 2 ml buffer solution to 50 ml with demineralised water in a 50 ml volumetric flask. The measuring solution is prepared by diluting 5 ml standard calcium chloride solution, 5 drops concentrated hydrochloric acid, and 2 ml buffer solution to 50 ml with demineralised water in a 50 ml volumetric flask. This solution is identical with the last one of the series of solutions serving in the construction of the calibration curve.

Hydrochloric acid, concentrated (specific gravity = 1.19 g/cm^3).

Cation exchange resin "Dowex 50" in the H cycle. 100 g resin are washed in the column with 5 M hydrochloric acid until the eluate contains no calcium when tested by the flame photometer. The resin is then washed thoroughly with demineralised water and allowed to dry in the air.

Perchloric acid, 72 %

Ammonium nitrate, reagent-grade, solid

Instruments

Wagner-type shaking machine, filter-type flame photometer with dispersion chamber (Zeiss Model III).

Conditions

Flame: acetylene-air
Wavelength measured: Ca 622 nm (filter Ca 63 J)
Concentration range: 0.5 to 5 g Ca/l
Method of evaluation: single calibration curve method.

Construction of the calibration curve

Volumes of 0, 1, 2, 3, 4 and 5 ml standard calcium chloride solution are measured by pipette into 50 ml volumetric flasks (i.e. 0, 1, 2, 3, 4 and 5 mg Ca), 5 drops conc. hydrochloric acid, 2 ml buffer solution are added to every flask, the solutions are made up to the mark with demineralised water and stirred. The solutions are then measured as described in the procedure. A calibration curve is then plotted from the measured values, galvanometer deflections being plotted versus the calcium concentration expressed in mg/50 ml.

Procedure

A 100 ml sample is heated with 50 ml hydrochloric acid until the dispersed substances have dissolved. After cooling, the solution is diluted with demineralised water to 1 litre in a volumetric flask and stirred. A 10 ml aliquot of this solution is measured by pipette into a 250 ml flask, diluted to about 150 ml and 1 g ion exchange resin is added. The flask is fixed to the shaking machine and agitated for 15 minutes. The solution is filtered with a loose paper filter: the resin particles must be transferred quantitatively to the filter. The resin is washed ten times with demineralised water on the filter. The paper is then dried in a drying box at 60 °C and the resin is transferred to a platinum dish. In this dish the resin is covered with 2.5 g solid ammonium nitrate and heated on an asbestos plate over a flame, until the initial reaction has taken place. The dish is then heated with the flame of a BUNSEN burner until all the organic substances have been burnt. After cooling, 5 ml perchloric acid is added and the mixture is heated on a sand bath until dry, but not longer. The residue is moistened with 5 drops conc. hydrochloric acid, 30 ml demineralised water are added, and the dish is heated until the residue has wholly dissolved. The solution is now transferred to a 50 ml volumetric flask, 2 ml buffer solution are added, the solution is made up to the mark with demineralised water and stirred. Meanwhile the photoelectric cell is allowed to "run in" by dispersing a solution containing roughly 3 mg Ca in 50 ml, the calcium filter being inserted. With the maximum galvanometer sensitivity, the zero point is set for the zero solution and the maximum galvanometer deflection is set for the measuring solution with the use of the diaphragm (a deflection of 250 divisions is suitable). The setting of the zero point and maximum deflection is repeated several times, until the two values agree. Immediately afterwards, the analysed solution is measured. If several solutions are to be measured, the maximum deviation must be checked before every measurement, while the zero point is checked after every three measurements. The calcium amount is then read off the calibration curve in mg/50 ml.

Calculation

$$\text{mg Ca/ml} = a \cdot 100$$

a being the amount of calcium read off the calibration curve, in mg/50 ml. The reproducibility of the determination is about ±15 % rel.

14.2.4 Determination of alkali metal oxides in tungstic acid [338]

Quality testing of technical tungstic acid also involves determinations of Na_2O and K_2O. These determinations are also prescribed by the Czechoslovak State Standard [151]. The method is based on extracting the sample with water, acidifying with hydrochloric acid, and determining the alkali in the extract. It was shown experimentally that the alkalis are digested quantitatively. The determination of potassium is unfavourably influenced by the presence of sodium. In order to eliminate this influence, standard solutions are prepared for the construction of the calibration curve so as to contain potassium and sodium in roughly that ratio contained by the samples to be analysed. Calcium interferes in the determination only if present in amounts larger than conventionally found in the samples.

Solutions

The stock potassium solution is prepared by dissolving 1.583 g KCl in a 1 litre volumetric flask, making up to the mark with demineralised water and stirring: 1 ml of this solution corresponds to 1 mg K_2O.

The stock sodium solution is prepared by dissolving 1.8857 g NaCl in a 1 litre volumetric flask, making up to the mark with demineralised water and stirring: 1 mg of this solution corresponds to 1 mg Na_2O. Both stock solutions are stored in polythene vessels.

Hydrochloric acid, concentrated (specific gravity = 1.19 g/cm^3)

Instrument

Photometer with dispersion chamber and filters (Zeiss Model III)

Conditions

Flame: acetylene-air

Wavelength measured: Na 589.3 nm, K 768.2 nm (Zeiss filters Na 59 and K 77 J)

Range of the determination: 0.01 to 0.025 % K_2O: 0.02 to 0.08 % Na_2O.

Method of evaluation: method of frames (see section 10.6.2)

Construction of the calibration curves

The calibration curves are constructed by measuring a series of standard solutions prepared according to Table 14.2. The volumes of stock solutions stated in Table 14.2 are measured by burette into volumetric flasks 1000 ml in volume, diluted with demineralised water up to the mark and stirred. The solutions are stored in bottles of polythene. The standard solutions are measured in the manner described in the procedure. The zero point of the galvanometer is set with demineralised water and the maximum deflection is set for solution No. 4 (Table 14.2). The maximum deflection is set by means of the diaphragm so as not to exceed a value of 400 scale divisions.

Table 14.2. COMPOSITION OF STANDARD SOLUTIONS

Notation	Amount of K_2O (related to a sample weight of 2 g), %	Amount of Na_2O (related to a sample weight of 2 g), %	Volume of K_2O stock solution, ml	Volume of Na_2O stock solution, ml
1	0.010	0.020	2	4
2	0.015	0.040	3	8
3	0.020	0.060	4	12
4	0.025	0.080	5	16

A calibration curve is constructed from the measured values by plotting the galvanometer deflections versus the K_2O and Na_2O concentrations, expressed directly in percentages of Na_2O and K_2O in the sample (with a sample weight of 2 g). Both calibration curves only serve for determining the approximate content of K_2O and Na_2O.

Procedure

A sample of 2 g is weighed with an accuracy of ±0.01 g. The sample is transferred quantitatively into a quartz flask 250 ml in volume, 80 ml demineralised water is added and the solution is acidified with 5 to 7 drops hydrochloric acid. A reflux condenser is fitted to the flask and the sample is boiled for 2 hours. The insoluble tungstic acid is then separated by filtering with a paper filter and washed on the filter with a small amount of acidified demineralised water (5 drops hydrochloric acid in 100 ml water). The filtrate is collected in a 100 ml flask. After making up to the mark with demineralised water and stirring, the filtrate is ready for measurement.

The photometer is allowed to "run in" for 10 to 20 minutes with the sample solution, the zero point is set for demineralised water and the maximum deviation is set for solution No. 4 (see Table 14.2). The maximum deviation is set by use of the diaphragm to the same values which were selected for the construction of the calibration curves. The sample solution is then measured. The measurement is repeated at least three times. The mean value of the measured results is used to read the approximate K_2O and Na_2O concentrations from the respective calibration curves. The next highest and lowest standard solutions and the sample are then measured quickly one after the other in order to avoid any change taking place in the measurement conditions. The content of the alkali oxides in the sample is calculated in relation to the standard solutions. Calculation:

$$\% \, K_2O \text{ (or } Na_2O) = \frac{V_3 - V_1}{V_2 - V_1} (M_2 - M_1) + M_1,$$

where M_1 is the amount of K_2O (or Na_2O) in the next lower standard solution, expressed as the percentage in the sample,
M_2 is the amount of K_2O (or Na_2O) in the next higher standard solution, expressed as the percentage in the sample,
V_1 is the galvanometer deflection corresponding to the next lower standard solution,
V_2 is the galvanometer deflection corresponding to the next higher standard solution,
V_3 is the galvanometer deflection corresponding to the sample solution.

The reproducibility of the determination is ± 15 rel. %. HEGEDÜS et al. [353] describe a different method of determining sodium and potassium in tungstic acid. To separate tungsten from the elements mentioned, these authors distil tungsten chlorides in a stream of CCl_4.

14.2.5 Determination of potassium in vanadium contact masses [220]

Vanadium contact masses are used in the production of sulphuric acid for catalytic oxidation of sulphur dioxide. One of the components which influence the properties of the contact mass is potassium [116]. Therefore, the potassium content must be determined in the production, and application of the contact mass. The method described permits a simple and rapid determination of potassium in this relatively complex material. The content of the other components of the contact mass is generally known approximately and, therefore, model solutions may be used to eliminate their influence.

Solutions and reagents

Sulphuric acid, concentrated (specific gravity = 1.84 g/cm^3)
Hydrofluoric acid, 40 %
Ammonium vanadate solution* containing 1 mg V in 1 ml: this is prepared by dissolving 1.1480 g NH_4VO_3 in water, making up to 500 ml with water and stirring. Iron sulphate solution containing 1 mg Fe in 1 ml: prepared by dissolving 2.4910 g $FeSO_4 \cdot 7\,H_2O$ in water acidified with several drops of dilute sulphuric acid, making up to 500 ml with water and stirring.
Aluminium sulphate solution*, containing 1 mg Al in 1 ml, prepared by dissolving 3.1705 g $Al_2(SO_4)_3$ in water acidified with several drops dilute sulphuric acid, making up to 500 ml with water and stirring.
Sodium sulphate solution, containing 1 mg Na in 1 ml, prepared by dissolving 1.5430 g Na_2SO_4 in water, making up to 500 ml and stirring.
Calcium sulphate solution, containing 0.5 mg Ca in 1 ml, prepared by mixing 0.6248 g $CaCO_3$ with 400 ml water and adding dilute sulphuric acid (1 to 2) dropwise with constant stirring, until all the carbonate has dissolved. The solution is then made up to 500 ml with water and stirred.
Sulphuric acid solution, prepared by diluting 100 ml conc. sulphuric acid to 1000 ml with water with stirring.
Potassium sulphate standard solution, containing 1 mg K/ml (Appendix II).
The blank solution is prepared by diluting 40 ml ammonium vanadate

---

* If the solutions are not clear, they must be filtered

solution, 25 ml ferrous sulphate solution, 5 ml aluminium sulphate solution, 2.5 ml sodium sulphate solution, 15 ml calcium sulphate solution, and 10 ml sulphuric acid solution to 500 ml. This solution is identical with the first solution of the calibration series. The measuring solution is prepared by diluting 40 ml ammonium vanadate solution, 25 ml ferrous sulphate solution, 5 ml aluminium sulphate solution, 2.5 ml sodium sulphate solution, 15 ml calcium sulphate solution, 10 ml sulphuric acid solution and 65 ml potassium sulphate standard solution to 500 ml. This solution is identical with the last one in the calibration series.

Instrument

Photometer with dispersion chamber and filters (Zeiss Model III)

Conditions

Flame: acetylene-air
Wavelength measured: K 768.2 nm (Zeiss filter K 77 J)
Concentration range: 1 to 15 % K.
Method of evaluation: single calibration curve method with the use of model standard solutions.

Construction of the calibration curve

Into 500 ml volumetric flasks, volumes of 0, 15, 27.5, 40, 57.5 and 75 ml standard potassium solution (i.e. 0, 3, 5.5, 8, 11.5 and 15 mg K/100 ml) are measured by burette. 40 ml ammonium vanadate solution, 25 ml ferrous sulphate solution, 5 ml aluminium sulphate solution, 2.5 ml sodium sulphate solution, 15 ml calcium sulphate solution and 10 ml sulphuric acid solution are added to every flask. All the solutions are made up to the mark, stirred and measured as described in the procedure. The measured galvanometer deflections are plotted versus the potassium concentration expressed in mg/100 ml. The calibration curve is best constructed anew for every series of samples.

Procedure

1 g of the well-ground sample is weighed into a platinum dish, moistened with several ml of water, 10 ml concentrated sulphuric acid and roughly 30 ml hydrofluoric acid are added, and the mixture is heated on a sand bath until white fumes appear. After cooling, the contents of

the dish are diluted with water, transferred to a 200 ml volumetric flask, made up to the mark with water and stirred. A 20 ml aliquot of this solution is pipetted into a 100 ml volumetric flask, diluted with water to the mark and stirred. This solution is used for measurement (after being filtered, if necessary). The photometer is allowed to "run in" with a medium-concentration standard solution (10 to 20 minutes) with the potassium filter inserted, then the zero point and the maximum galvanometer deflection are set with the zero and measuring solutions, respectively. Before measurement of the sample solution, the maximum deflection is set to the same value which was used for measurements of the calibration series. Now the sample solution is measured. The measurement is repeated three times, and the mean value is used to read the amount of potassium off the calibration curve.
Calculation:

$$\%\mathrm{K} = a$$

where a is the amount of potassium in mg/100 ml, read off the calibration curve. The reproducibility of the determination is about ± 0.5 absol. %.

14.2.6 Other possible applications of flame photometry in the chemical industry

Among other examples, let us mention the determination of sodium in salts of organic acids [446], determination of sodium in sulphuric acid [512], determination of sodium, potassium, and lithium in alkali metal salts [413], determination of sodium and lithium in beryllium solutions [536], determination of sodium and potassium in calcium carbonate and phosphate [627], determination of sodium in aluminium oxide and its hydrate [91], [350], determination of sodium in potassium carbonate, hydroxide, and chloride [792], determination of sodium [599], lithium [599], strontium [599], barium [599], [613], [155] and lanthanum [600] in uranium, determination of potassium in the ammonia synthesis catalyst [522], determination of sulphates in reagents [122], determination of tetraethyl lead in gasoline [314], [421], of tributylphosphate in extraction gasoline [95], [491], [492], determination of boron in organic substances [103], determination of lithium, sodium, potassium and calcium in graphite [40] and of calcium in cellulose [569]. The application of flame photometry in the analysis of oils [147], in the rubber industry [753] and in fuel analysis [53], [227], [407], [422], [427], [621], [685] are also worth mentioning.

In some cases, the flame photometer may be used with advantage to determine some technological parameters. For example, it makes it possible to determine the time for which materials are delayed in reactors in a rapid and simple manner. At a certain moment, a suitable substance not originally contained in the material is added to it at the inlet to the reactor or other piece of equipment: a substance easily determined by flame-photometric means should be selected, e.g. Na, Li, K. Samples are then taken at specified intervals of time at the outlet from the equipment in question, and the substance added is determined in these samples. Since a relative content of the substance added in the individual samples is involved, there is no need to use a calibration curve and it suffices to record the galvanometer deflections. The deflections corresponding to the individual samples are plotted on a graph versus the time at which the samples were taken. From this graph, which is similar in shape to the Gaussian curve, the rate of passage of the material and its scatter in the course of passage may be read. In this way, the time of delay and scatter of material in a rotary furnace were determined [220].

14.3 Analysis of silicates, minerals, ores and rocks

Materials included in this group are of a greatly varied composition, the main component, however, being silicic acid, so that the method of decomposition is similar for all samples (except sulphide ores). This group also includes materials of technical importance, like glass, refractories, cement, etc. Silicate materials usually are decomposed by hydrofluoric acid, mixed with other acids like sulphuric or perchloric acid. Alkalis may also be decomposed by the LAWRENCE–SMITH method (sintering the sample with calcium carbonate), but generally the former procedure is preferred. Methods may differ for the individual materials, and they also vary according to the content of interfering elements.

To show how flame photometry simplifies determinations which are very difficult when classic methods are used, we describe the determination of alkalis, including rubidium and cesium, in silicates, and the determination of magnesium and sodium by means of atomic absorption. The reader will find other procedures in the literature (see citations listed in Appendix III).

14.3.1 Determination of lithium, sodium, and potassium in silicates

This method is suitable for the analysis of conventional silicate rocks, minerals, glass, ceramic materials, or of other raw materials provided they do not contain too much calcium, and also for the determination of lithium and strontium. The radiative interference of calcium and strontium is suppressed by the addition of ammonium phosphate, which forms non-volatile compounds with alkaline earth elements. The interfering ionisation effect is eliminated in the determination of sodium and lithium by the addition of potassium chloride. In the determination of potassium the analogous effect of sodium is neglected, a circumstance which is permissible owing to the ratio of these elements in conventional samples. The effect of sodium need be considered only if the excess of sodium over potassium is more than an order of magnitude, e.g. in some sodium glass types.

Solutions and reagents

Perchloric acid, 72 %
Hydrofluoric acid, 40 %
Hydrochloric acid (specific gravity = 1.19 g/cm^3)
Boric acid, saturated aqueous solution.

Additional solution for the determination of lithium and sodium: this is prepared by dissolving 6 g potassium chloride, 100 g ammonium dihydrogenphosphate, and 250 ml hydrochloric acid in distilled water, and making up to 1000 ml with water.

Additional solution for the determination of potassium: this solution is prepared in the same way, but without the addition of potassium chloride.

Standard solutions of lithium, potassium and sodium chlorides are prepared according to data given in Appendix II. These solutions contain, in 1 ml, 1 mg sodium, 1 mg potassium, and 0.1 mg lithium.

Instrument

Flame photometer with spray chamber and filters (Zeiss Model III).

Conditions

Flame: acetylene-air
Wavelengths measured: Li 671 nm, Na 589 nm, K 768 nm
Method of evaluation: calibration curve method

Construction of calibration curves:

To construct the calibration curves, two series of calibration solutions are prepared, one for lithium and sodium and one for potassium. Volumes of these solutions as given in Table 14.3 are transferred by pipette into 50 ml volumetric flasks. The solutions are made up to the mark with water, mixed, and measured as described as the procedure. The measured galvanometer deviations are plotted versus concentration of the respective element in the solution. Three calibration curves are thus obtained: one for lithium, the second for sodium, the third for potassium.

Procedure

A 1 g sample is decomposed in a platinum dish with a mixture of 10 ml hydrofluoric acid and 1 ml perchloric acid. The sample is mixed well with the acids and the mixture is heated on a sand bath till white perchloric acid fumes appear. 5 ml hydrofluoric acid are again added, the mixture is stirred and evaporated till dry. The residue is sprayed with 1 ml perchloric acid and 5 ml saturated boric acid solution and evaporated till entirely dry to remove fluoride residues. The residue is dissolved by digesting with 10 ml water. 4 ml concentrated hydrochloric acid are added and the mixture is heated till everything is dissolved. After cooling, the solution is transferred into a 100 ml volumetric flask, made up to measure, and mixed.* A 10 ml aliquot of this stock solution is transferred by pipette into a 50 ml volumetric flask for the determination of lithium and sodium, and into a second flask for the determination of potassium. These solutions are now ready for measurement.

The photometer is allowed to "run in" with a medium concentration calibration solution, with use of an appropriate filter. The zero point

* In the case of silicates difficult to decompose, the decomposition with hydrofluoric acid must be repeated several times. Even then some minerals may remain non-decomposed: these, however, are usually types which do not contain alkali metals, e.g. topaz.

Table 14.3. PREPARATION AND COMPOSITION OF STANDARD SOLUTIONS

Notation of standard solutions	Li		Na		K		Additive solution
	ml	mg/50 ml	ml	mg/50 ml	ml	mg/50 ml	ml
1	0.5	0.05	0.5	0.5			10
2	1.0	0.1	1	1			10
3	1.5	0.15	2	2			10
4	2.0	0.2	3	3			10
5	4.0	0.4	4	4			10
6	6.0	0.6	5	5			10
7	8.0	0.8	6	6			10
8	10.0	1.0	7	7			10
K_1					0.5	0.5	10
K_2					1	1	10
K_3					2	2	10
K_4					3	3	10
K_5					4	4	10
K_6					5	5	10
K_7					6	6	10
K_8					7	7	10

is then adjusted and the maximum galvanometer deviation is set for the blank or the measured solution. Finally, the calibration curve and the sample solutions are measured. The lithium, sodium, and potassium content is read from the respective calibration graphs from the recorded galvanometer deviations.

The procedure described is suitable for amounts of up to 1 % lithium or up to 7 % sodium and potassium in the sample. For lower amounts we select a lower concentration range of the calibration solutions, or we take larger volumes of the stock solution.

14.3.2 Determination of rubidium and caesium [543]

Materials analysed are mainly rocks (granites, greisens) and sometimes minerals (feldspars, micas). The rubidium content is of the order of hundredths of a percent, exceptionally up to 1 %. The caesium content is usually one order lower. In all these materials, potassium is present at the same time in an excess of several orders. Therefore, to isolate the rubidium and caesium lines from the doublet K 770 nm we need a monochromator. To determine rubidium, the line 794 nm is used, while caesium is determined with the help of the Cs 852 nm line. Since the caesium determination is affected less by the radiation interference of potassium, a larger monochromator slit is used.

Potassium interferes most, since on the one hand it influences the ionisation equilibrium of rubidium and caesium in the flame, and on the other causes strong radiation interference. In order to eliminate its influence, we first determine potassium and adjust its concentration in the solution measured to 300 mg/100 ml. Sodium does not interfere in amounts up to 100 mg/100 ml (10 % in the initial sample). Caesium does not affect the determination of rubidium in amounts up to 60 mg//100 ml (6 % in the initial sample), though in the case of caesium there is a slight increase in the limit of experimental error. The cause is the radiation interference of CaO bands with edges at 824 and 872 nm. This would have to be considered when analysing samples with high calcium content. Aluminium has no influence in amounts up to 150 mg/100 ml (15 % in the original sample).

Solutions

The additive potassium solution contains 50 g K/l. It is prepared by dissolving 118 g K_2SO_4 and making up to 1000 ml. A standard solution of potassium, rubidium, and caesium, all in cocentrations of 1 mg/ml, is prepared according to Appendix II. The caesium solution is then diluted 10 times (10 ml into a 100 ml volumetric flask). Preparation of calibration solutions: volumes of 2.5, 2, 1.5, 1 and 0.5 ml standard rubidium solution, and dilute caesium solution, are transferred by pipette into a 50 ml volumetric flask. 1 ml concentrated hydrochloric acid and 3 ml additive potassium solution are added, and the solution is made up to measure and mixed. The solution for zero setting contains 300 mg K/100 ml. It is prepared by making up 6 ml K_2SO_4 solution and 2 ml concentrated hydrochloric acid to 100 ml.

Instrument

Zeiss Model SPM 1 monochromator with photomultiplier sensitive in the red region (Type FEU 22) on the outlet slit; atomiser with cloud chamber.

Flame: acetylene – air
Evaluation: by means of the calibration curve method.

Procedure

A 1 g sample is decomposed by heating with hydrofluoric and perchloric acid (see 14.3.1). When the sample has been dissolved, however, it is transferred quantitatively into a 50 ml volumetric flask, mixed, and made up to the mark. A 5 ml aliquot of the clear solution is transferred into a 50 ml volumetric flask for the determination of potassium, and another 25 ml into a 50 ml flask for the determination of rubidium and caesium. Potassium is determined first.

Determination of rubidium and caesium: To 25 ml of the solution in a 50 ml volumetric flask add 1 ml concentrated hydrochloric acid and enough potassium sulphate solution to give a final concentration of 150 mg K in 50 ml. After shaking, the measurement is carried out under conditions given in Table 14.4. The calibration curves are linear and do not pass through the origin, but instead through a point corresponding to the radiation interference of potassium, which we determine from the standard zero setting. The precision of the determination is ±6 % of the result on a 95 % significance level.

Table 14.4. EXPERIMENTAL CONDITIONS FOR THE DETERMINATION OF RUBIDIUM AND CAESIUM

	Rubidium	Caesium
Wavelength measured, nm	794	852
Slit width, mm	0.10	0.15
Highest standard, mg/50 ml	2.50	0.25

14.3.3 Determination of magnesium

Magnesium is one of the elements which is suitable for determination by means of atomic absorption. The Mg 285 nm resonance line lies in hte ultraviolet spectral region, where the radiation of the flame is very slight. Interference in the magnesium determination is due only to the formation of non-volatile compounds, mainly with aluminium, sometimes with silicon. Sulphate and phosphate ions interfere in the cooler part of a natural gas–air flame. This interference may be eliminated by calcium and by oxine.

Solutions

Magnesium standard solution: 0.1 g pure magnesium is dissolved in the necessary amount of hydrochloric acid (1 to 4) and, after cooling, made up to 100 ml. This solution is then diluted twentyfold. The final solution contains 0.05 mg Mg in 1 ml. Calcium solution: 5.9 g $Ca(NO_3)_2 \cdot 4\,H_2O$ are dissolved in water and made up to 100 ml. The solution contains 10 mg Ca/ml. Other calcium salts may also be used, but they may not contain more magnesium than 0.001 % of the calcium content.

Oxine solution: 50 g oxine are dissolved in methyl alcohol and made up to 500 ml with alcohol. The resulting solution is of 10 % strength. Hydrochloric acid, doubly-distilled (azeotropic mixture).

Calibration solutions: 1 ml Ca solution, 5 ml oxine, and in turn 10, 6, 2, 1, and 0.5 ml Mg solution are measured into 50 ml flasks and made up to measure. The calibration solutions contain 0.5, 0.3, 0.1, 0.05 and 0.025 mg magnesium in 50 ml.

Instrument

Monochromator with photomultiplier tube on the outlet slit (Types FEU 18 or 1P28). The radiation source is a magnesium hollow cathode discharge lamp, fed with a current of 10 to 14 mA. The atomiser and burner are taken from the equipment of the Zeiss flame photometer.
Flame: acetylene–air.
Line measured: Mg 285 nm.
Concentration range: 0.025 to 0.5 mg Mg.
Evaluation: The calibration curve method is used. When the zero solution is atomised, the full deviation of the measuring instrument, I_0, is adjusted by altering the amplification or the slit width of the monochromator. The standards are measured one by one in the ascending order, and the respective deviations I are read. The calibration curve is constructed by plotting absorbance values $A = \log I_0/I$ against the magnesium

concentration. The dependence is linear, passing through the origin. From the absorbance values corresponding to the analysed samples, their concentraions are read from the calibration curve.

Procedure

From the stock solution obtained by decomposition of the sample with hydrofluoric and perchloric acid for the determination of alkali metals (see 14.3.1), a 25 ml aliquot is measured into a 50 ml volumetric flask, 5 ml oxine solution are added, and 1 ml calcium solution. The mixture is made up to volume and mixed. This solution is then ready for measurement.*

The magnesium content is calculated according to the formula

$$\% \text{ Mg} = X/2.5,$$

where X is the magnesium content found in milligrammes. Magnesium may thus be determined in the concentration range 0.01 to 0.2 % Mg in the solid sample. If the magnesium content is greater, a correspondingly smaller volume of the stock solution is taken. The precision of the determination is ±3 % of the result.

14.3.4 Determination of sodium in limestone by means of atomic absorption

The determination of alkali metals in limestone is important from the point of view of the technology of cement production, where an increased content of alkali metals causes agglomeration. In the determination by the emission method, the sodium line at 589 nm is subject to interference by the molecular spectrum of calcium. This radiative interference may be suppressed in part by the addition of phosphate ions. When

* In the analysis of limestone the sample is dissolved in hydrochloric acid. Hydrofluoric acid is only used to decompose the insoluble residue. The solution of this decomposed residue then is added to the filtrate obtained from the hydrochloric acid treatment, and the combined solution, made up with water to a given volume, is used for the analysis. In this case, oxine and calcium solutions are not added, but the calibration solutions must contain approximately the same amount of calcium as the analysed sample solutions. If a magnesium-free calcium salt is not available, the magnesium content can be determined by the standard addition method.

sodium is determined by means of the atomic absorption method, the radiative interference of calcium is fully eliminated.

Solutions

Standard sodium solution: This is prepared by dissolving 0.2540 g sodium chloride in water and making up to 200 ml. The solution contains 0.25 mg Na in 1 ml. The solution is stored in a polythene bottle.

Standard calcium solution: This is prepared by dissolving 50 g calcium carbonate in hydrochloric acid in a 500 ml volumetric flask and making up to the mark with the acid. 1 ml of this solution corresponds to 100 mg $CaCO_3$. The calcium carbonate employed must be free of sodium. If this is not the case, its sodium content may be determined by means of the standard addition technique.

Blank solution: 10 ml standard calcium solution is measured by pipette into a 100 ml volumetric flask and made up to the mark with distilled water.

Instrument

The atomic absorption instrument consists of a monochromator, dispersion device with mist chamber, burner for a town gas-air flame, and galvanometer, all belonging to the Model III Zeiss flame photometer. The monochromatic radiation source is a sodium discharge lamp with damping coil, the type used for example in polarimetric measurements. A photoelectric multiplier tube is located at the monochromator exit slit. The current formed in the photomultiplier tube is measured with a mirror galvanometer. Instead of a monochromator, a sodium interference filter may also be used, and the detector may be a selenium barrier photocell.

Conditions

The town gas – air flame is adjusted so as to make its blue cone about 1 cm in height and to let the radiation of the sodium lamp pass above the blue cone. The sodium doublet at a wavelength of 589 nm is measured.

Construction of calibration curves: Volumes of 1, 2, 4, 6, 8 and 10 ml standard sodium solution and 5 ml standard calcium solution are measured by pipette into 50 ml volumetric flasks. Standards thus prepared contain 0.25, 0.5, 1.0, 1.5, 2 and 2.5 mg Na in 50 ml. The solutions are made up to the mark with distilled water and stirred. The solutions are measured in the manner described in the procedure. The calibration curve should pass through the origin of the coordinate system. If this is not the case, the calcium salt employed is contaminated with sodium. With the higher standards the calibration curve bends somewhat towards the abscissa, which is caused by the radiation of the flame itself.

Procedure

A sample of 0.5 g is weighed and dissolved in 5 ml hydrochloric acid. After cooling the solution is transferred into a 50 ml volumetric flask and made up to the mark with water. This solution is now ready for measurement. The sensitivity of the instrument is set by means of the diaphragm or slit so as to give a maximum deviation (I_0) when the zero solution is dispersed. Now the deflections corresponding to the individual standards are measured. Absorbance values $A = \log I_0/I$ are plotted versus concentration. From the absorbance values of the individual samples, the respective concentrations are read off the calibration curve.

The sodium concentration is calculated from the relation

$$\%\,\mathrm{Na} = \frac{a}{5}$$

where a is the amount of sodium in mg/50 ml, read off the calibration curve. The reproducibility of the determination is about 3 % rel.

If the samples analysed have a large insoluble residue, in which the sodium content must also be determined, the sample is filtered after dissolution in hydrochloric acid, the precipitate on the filter is washed thoroughly, the filter is ignited in a platinum crucible, and the ash is volatilised with hydrofluoric and perchloric acids. After dissolution in dilute hydrochloric acid, the solution is added to the first filtrate and made up to the mark with water.

The above procedure permits amounts of 0.05 to 0.5 % Na to be determined. If the sensitivity is to be increased, the butterfly-type burner is turned into the direction of the optical axis. This increases the sensitivity roughly 5 times.

14.3.5 Other possible applications

Flame photometry has been applied to the determination of alkali metals in glass [70], [96], [141], [265], [341], [352], [355], [357], [359], [360–363], [474], [584], [604], [651], [653], [766], [801], [806], in minerals and ores [37], [70], [81], [99], [148], [149], [168], [181], [228], [266], [332], [333], [384], [406], [429], [445], [467], [544], [556], [584], [588], [665], [669], [746], [782], [803], [806], in ceramic materials [29], [145], [188], [235], [266], [311], [563], [630], [673], in portland cement [31], [46], [189], [235], [462], [515], [563], to the determination of calcium in glass [133], [384], [386], [445], [467], [544], [556], [584], [588], [655], [669], [746], [782], [803], [806], in minerals and ores [29], [80], [451], [455], [581], [631], [731] and in slag [396], [653], [727]. Furthermore, for the determination of strontium in minerals and ores [275], [331], [458], [581], [762] and in portland cement [190], determination of magnesium in glass [653] in minerals and ores [80], [251], [495], in portland cement [140], in slag [2], [544], [653]. Furthermore, it is worthwhile mentioning the determination of manganese [196], lanthanum [400], thallium [838], aluminium [184], iron [173], and boron [171] in minerals and ores. Determinations of manganese [653], lanthanum [400], and aluminium in glass and of manganese in portland cement are also described in the literature.

14.4 Applications of flame photometry in agricultural chemistry

Agrochemical analyses mainly involve the determination of potassium, which is one of the basic nutrients. However, the use of flame photometry is not limited to the determination of this one element. The materials analysed are greatly varied. They include industrial and natural fertilisers, soil samples, plants, etc. Each type of material demands a different procedure for sample preparation. This is specially important in agrochemical analyses. For example, the procedure for preparing extracts of natural fertilisers or soils is an integral part of the analysis. Since these extraction procedures are usually standard ones, based on empirical experience, they must be adhered to exactly in order to obtain comparable results. These problems belong to the sphere of agricultural analysis and have no immediate relation to the methods of flame photometry. The methods of sample preparation are discussed in detail in the respective agrochemical monographs and handbooks [387], [525], [576], [632].

As an example of the application of flame photometry in agricultural analysis, let us mention the determination of potassium in a lactate extract: other possible applications will only be mentioned by reference to the original literature.

14.4.1 Determination of potassium in lactate extracts of soil

This method is used to determine that amount of potassium which is available to plants in the soil. It is based on extracting the sample with calcium lactate solution at a pH value of 3.6. Calcium, which would interfere in the determination of potassium, is precipitated from the solution with oxalic acid.

Solutions

A stock solution of calcium lactate is prepared by dissolving 308.23 g calcium lactate $(CH_3COHCOO)_2Ca \cdot 5\,H_2O$ in 2 litres distilled water, the solution being heated to speed up the dissolution process. 200 ml 5N hydrochloric acid solution is then added and after cooling the solution is made up to 4 l with distilled water: the pH of this solution should be 3.2. A dilute calcium lactate solution is prepared by diluting 500 ml of the stock solution to 10 l with distilled water: the pH of this solution should be 3.6.

A potassium chloride stock solution is prepared by dissolving 1.5830 g potassium chloride in distilled water and making up to 1 litre: 1 ml of this solution contains 1 mg K_2O. The potassium chloride standard solution is prepared by diluting 100 ml of the stock solution to 1 litre with distilled water. The zero solution is identical with the first solution of the calibration series, being prepared in the same manner.
The measuring solution is identical with the last solution of the calibration series, being prepared in the same manner.
Oxalic acid solution, 10 %.

Instruments

Wagner-type shaking machine, flame photometer with dispersion chamber fitted with filters (Zeiss Model III)

Conditions

Flame: acetylene – air
Wavelength measured: K 768.2 nm (filter Zeiss K 77 J)
Concentration range: 5 to 60 mg K_2O/100 g soil
Method of evaluation: single calibration curve method
Construction of the calibration curve

Volumes of 0, 10, 20, 30, 40, 50 and 60 ml potassium chloride standard solution (corresponding to 0, 10, 20, 30, 40, 50 and 60 mg K_2O/100 g soil) are measured by pipette into 500 ml volumetric flasks. 25 ml calcium lactate stock solution is then measured by pipette into each flask, the solutions are made up to the mark with distilled water and stirred. 25 ml of each solution is pipetted into a beaker, 2 ml oxalic acid solution is added, the mixture is stirred and allowed to settle overnight. The clear liqud is then decanted from each beaker, to be measured as described in the procedure. The galvanometer deflections are plotted versus the potassium concentration expressed in mg/100 g soil.

Procedure

An air-dried, finely ground sample of 5 g is agitated for 1.5 hours with 250 ml dilute calcium lactate solution in a flask fixed in a Wagner-type shaking machine. The solution is filtered with a fluted paper filter into a dry conical flask, the first filtrate portions being discarded. From the remainder, a 25 ml aliquot is measured by pipette into a beaker, 2 ml oxalic acid solution is added, the mixture is stirred and the precipitate formed is allowed to settle overnight. The clear portion of the solution is decanted to be used for the measurement. After the instrument has "run in" the zero position is set for the zero solution and the maximum deflection is set for the measuring solution. The sample solution is then measured, and the K_2O content is read off the calibration curve directly in mg/100 g soil. The reproducibility of the method is 10 % rel.

14.4.2 Other applications

The literature includes a large number of studies dealing with the application of flame photometry in agricultural chemistry, as this was the field in which the method first came to be applied fully. Most of these studies deal with soil analysis [5], [33], [57], [113], [134], [258], [291],

[307], [402], [403], [432], [434], [450], [457], [518], [519], [520], [527], [549], [575], [612], [695], [667], [671], [752], [756], [798] or analysis of plant material [5], [61], [83], [117], [159], [267–274], [305], [306], [393], [416], [433], [434], [435], [436], [449], [450], [482], [519], [551], [553], [612], [670], [672], [680], [700], [732].

14.5 Applications of flame photometry in biochemistry

Flame photometry is frequently employed as an analytical technique in clinical and hospital laboratories [501], [749] for diagnostic purposes. The materials analysed may be greatly varied. Most often, Na, K, and Ca are determined in blood serum, urine, etc., but there are many more possibilities. As an example, let us mention the determination of calcium in blood serum and in urine: other applications of flame photometry in biochemistry are listed in Appendix III.

14.5.1 Determination of calcium in serum and urine

The direct determination of calcium in the materials mentioned is highly unreliable, since the samples contain various interfering components, mainly phosphorus bound in different ways, but also sodium in variable concentrations and ratios; there are also great differences in the physical properties of the solutions. The determination of calcium in serum is also made difficult by the fact that samples available are usually quite small. Calcium is measured at a wavelength of 422.7 nm, where the radiative interference is minimum. Other interfering influences may be eliminated by the use of the method of standard additions. Since the relation between the galvanometer deflection and calcium concentration is linear in the range of 0 to 1 mg/100 ml, the calcium content in the sample may be determined by calculation.

Solutions

The calcium stock solution is prepared by weighing 0.4994 g $CaCO_3$, dissolving in a small amount of dilute nitric acid and making up to 1000 ml with water: 1 ml of this solution contains 0.2 mg calcium. For setting the sensitivity of the instrument this solution is diluted 100 times with water. Demineralized water is used throughout.

Instrument

Unicam SP 900 flame photometer, slit 0.08 mm

Conditions

Flame: acetylene – air
Wavelength measured: 422.7 nm; the background is measured at wavelengths of 418 and 428 nm.
Method of evaluation: standard addition method (see section 10.7.4).

Procedure

1 ml blood serum or urine is diluted with demineralised water to 50 ml (calcium content = x). A 10 ml aliquot of this solution is taken and 0.1 ml calcium stock solution is added with a pipette graduated in divisions of 0.01 ml (calcium content = $x + 2$). The instrument is set with the use of the 100 times diluted stock solution so as to make the galvanometer deflection at 422.7 nm equal to about 40 divisions. During the measurement, the same solution is used to check the intrument: moreover, the zero point must be checked before every measurement. This setting is done with water. The galvanometer deflection for the sample solution, V_1, is measured, the deflection for the sample solution with calcium solution added, V_2, and the spectral background is measured at 418 and 428 nm. The mean, V_p, of the two latter measurements is taken.

Calculation

$$\text{mg Ca/100 ml} = \frac{(V_1 - V_p) \cdot 10}{V_2 - V_1}.$$

14.6 Applications of flame photometry in other fields

Flame photometric methods have been utilised for the solution of analytical problems in a number of other fields of science and technology, especially in those cases where a large number of samples must be analysed. Besides the fields already discussed, flame photometry has been applied in metallurgy, where a number of emission and absorption methods are used in the analysis of light metals, light-weight alloys, steel, and a number of noble metals. Other applications have been in the electronics industry for the analysis of various materials used in making electron tubes and transistors, in the pharmaceutical industry for testing initial materials and final products, etc.

Automatic analysers for the determination of sodium [557], and for determining the purity of coke furnace gas [729], as well as continuous analysers for testing feed water [836], have been developed on the basis of flame photometry.

Since the scope of this book does not permit these methods to be described even briefly, we are presenting a review of the literature dealing with applications of flame photometry, in Appendix III: this permits the reader to orient himself rapidly.

Appendix I. TABLE OF SPECTRAL LINES AND BANDS

Wavelength nm	Character of radiation	Element	Wavelength nm	Character of radiation	Element
213.9	l	Zn	374.9	d	Fe
228.8	l	Cd	377.6	l	Tl
253.7	l	Hg	381	p	Mg
267.6	l	Au	383	p	Mg
285.2	l	Mg	385	p	C_2
306	p	OH	386.0	l	Fe
324.8	l	Cu	387.4	d	Co
326.1	l	Cd	388	p	Mg
327.4	l	Cu	389.5	d	Co
328.1	l	Ag	391	p	Mg
303.3	d	Na	398.8	l	Yb
338.3	l	Ag	403.1	l	Mn
340.5	l	Pd	403.3	l	Ga
341.3	l	Co	403.4	d	Mn
341.5	l	Ni	404.4	l	K
343.5	l	Rh	404.7	l	K
345.4	d	Co	405.8	l	Pb
346.0	d	Ni	407.8	i	Sr
349.3	l	Ni	410	p	C_2
350.2	l	Co	410.2	l	In
351.2	d	Co	417.2	l	Ga
351.5	l	Ni	420.2	l	Rb
352.5	l	Ni	421.6	i	Sr
352.7	l	Co	421.6	l	Rb
352.9	d	Co	422.7	l	Ca
356.6	l	Ni	425.4	l	Cr

Appendix I Continued

Wavelength nm	Character of radiation	Element	Wavelength nm	Character of radiation	Element
356.9	l	Co	427.5	l	Cr
357.5	l	Co	429.0	l	Cr
357.9	l	Cr	438	p	La
359.4	l	Cr	442	p	La
360.5	l	Cr	451.1	l	In
361.0	l	Pd	455.5	l	Cs
361.1	l	Ni	459.4	l	Eu
361.9	l	Ni	462	p	Gd
362	p	Mg	462.7	l	Eu
363.5	l	Pd	467	p	Lu
365.8	l	Rh	472	p	B
368.4	l	Pb	478	p	Yb
369.2	l	Rh	481	p	Ti
370.0	p	Mg	482	p	Y
372.0	l	Fe	483	p	Tm
372.8	l	Ru	485	p	Y
373.5	l	Fe	485	p	Ti
373.7	l	Fe	485	p	Yb
374.6	l	Fe	486	p	Sc
488	p	Ba	546	p	Er
491	p	Tm	547	p	V
493.4	i	Ba	548	p	B
494	p	B	549	p	Ba
496	p	Ti	549	p	Dy
497	p	Ba	552	p	Er
498	p	Yb	553	p	Tm
500	p	Ti	553.6	l	Ba
502	p	Ba	554	p	Ca
504	p	Er	555	p	Yb
506	p	V	555.7	l	Yb
511	p	Ho	556	p	Cr
512	p	Al	559	p	Mn
513	p	Ba	560	p	Er
516	p	Ho	560	p	La
516	p	Dy	561	p	Mn
517	p	Ti	561	p	Fe

Appendix I Continued

Wavelength nm	Character of radiation	Element	Wavelength nm	Character of radiation	Element
517	p	Lu	562	p	Cr
517	p	Yb	563	p	La
518	p	B	564	p	Tb
519	p	Mn	564	p	Ba
520	p	Dy	565	p	Fe
520.6	t	Cr	566	p	Ho
523	p	Mn	566	p	Er
523	p	V	568.3	l	Na
524	p	Ba	568.8	l	Na
526	p	Dy	569	p	Pr
527	p	Ho	570	p	Gd
528	p	V	570	p	Ba
532	p	Ho	572	p	Ca
533	p	Yb	572	p	Yb
534	p	Tb	573	p	Dy
535	p	Ba	573	p	Tb
535.1	l	Tl	574	p	V
536	p	Cr	576	p	Ti
536	p	Mn	577	p	Sc
538	p	Tm	579	p	B
539	p	Mn	579	p	Fe
540	p	Dy	581	p	Gd
542	p	Tm	581	p	Sc
542	p	Cr	582	p	Fe
542	p	Mn	583	p	Dy
543	p	La	585	p	Sc
544	p	Tb	585	p	Ho
544	p	Yb	585	p	Cr
545	p	Ti	586	p	Mn
587	p	Ba	615	p	Sc
587	p	Fe	615	p	Mn
587	p	Sm	618	p	Mn
587	p	Yb	619	p	Sc
588	p	Mn	622	p	Ca
589	p	Sc	622	p	Gd
589.0	l	Na	623	p	Eu

Appendix I Continued

Wavelength nm	Character of radiation	Element	Wavelength nm	Character of radiation	Element
589.6	l	Na	624	p	Sm
591	p	Gd	644	p	Ca
592	p	Tb	645	p	Sr
593	p	Sc	647	p	Eu
595	p	Sm	652	p	Sm
595	p	Sr	659	p	Sr
597	p	Sr	666	p	Sr
598	p	Eu	670.8	l	Li
598	p	Tb	680	p	Sr
599	p	Gd	702	p	Eu
599	p	Nd	702	p	Nd
600	p	Lu	704	p	Sr
602	p	Sc	712	p	Nd
601.8	l	Eu	715	p	Ti
602	p	Ca	743	p	La
603	p	Sm	766.5	l	K
604	p	Sc	769.9	l	K
604	p	Ba	780.0	l	Rb
605	p	Sr	792	p	La
605	p	Sr	794.8	l	Rb
607	p	Sc	818.3	l	Na
608	p	Tb	819.5	l	Na
610.4	l	Li	830	p	Ba
611	p	Sc	852.1	l	Cs
612	p	Gd	873	p	Ba
614	p	Sm	894.4	l	Cs
615	p	Y			

l – line corresponding to neutral atom, d – doublet, t – triplet, i – ionic line, p – molecular band

Appendix II. AMOUNTS OF COMPOUNDS TO BE WEIGHED IN PREPARING SOME FREQUENTLY EMPLOYED STANDARD SOLUTIONS

By dissolving the amount given in this table and making up to 1 litre with water, a solution containing 1 mg of the respective element in 1 ml is obtained

Element	Compound	Amount weighed g
Na	NaCl	2.5422
	Na_2SO_4	3.0892
K	KCl	1.9068
	K_2SO_4	2.2285
Li	LiCl	6.1091
	Li_2SO_4	7.9212
	$LiNO_3$	9,9350
	Li_2CO_3*	5.3235
Rb	RbCl	1.4148
	Rb_2SO_4	1.5619
Cs	CsCl	1.2668
	Cs_2SO_4	1.3614
Ca	$CaCO_3$*	2.4972
Ba	$BaCl_2 \cdot 6\,H_2O$	1.7786
	$BaCO_3$*	1.4369
	$Ba(NO_3)_2$	1.9029
Sr	$CrCl_2 \cdot 6\,H_2O$	3.0428
	$SrCO_3$*	1.6848
	$Sr(NO_3)_2 \cdot 4\,H_2O$	3.2376
Mg	MgO*	1.6580

* These substances are dissolved in a small excess of hydrochloric acid (1 to 1) and the solution is evaporated on a water bath until dry in order to remove the excess acid. The residue is dissolved in water and made up to one litre.

Appendix III. CITATIONS OF LITERATURE DESCRIBING ANALYTICAL METHODS

Element analysed	Material analysed	Literature	
		Emission	Absorption
Li	Alkali metal salts	413	
	Aluminium and its alloys	571	54
	Beryllium solutions	536	
	Biological material	43	86
	Ferrites	79	
	Graphite	40	
	Li–Mg alloys	640	
	Mother liquors of salts	291	
	Selenium	681	
	Silicates, ores, minerals	99, 181, 228, 237, 366, 384, 386, 429, 445, 476, 544, 580, 588, 746	844
	Glass		417
	Iron	460	
	Water	300, 320, 132	111
	Uranium	599	
Na	Alkaline earth carbonates	35	
	Alkaline metal salts	203, 413, 757, 792	
	Aluminium and its alloys	385, 528, 541, 793	
	Aluminium oxide and hydrate	91, 351	
	Ashes of various fuels	245	
	Beryllium solutions	536	
	Biological material	198, 718, 772	369, 652, 811
	Cellulose and paper	463	
	Cement		722, 748
	Calcium and its compounds	1, 627	
	Dough and rice	399	
	Beer		280
	Foods	745	
	Graphite	40	
	Fruit		75
	Lead and its compounds	282	
	Limestone		656
	Phosphate-type luminescent materials		567

Appendix III Continued

Element analysed	Material analysed	Literature	
		Emission	Absorption
	Selenium	681	
	Salts of organic acids	446	
	Slag	499, 653	
	Na–K alloys	208	
	Oils	751	720
	Glass and ceramic materials	35, 246, 359, 653	417
	Silicates, ores, minerals	28, 148, 149, 168, 181, 224, 311, 445, 665, 669, 744, 782, 803	56, 200, 608
	Sulphuric acid	512	
	Soil	33, 57, 134	161
	Uranium and its compounds	599, 846, 849	
	Uranium ores and concentrates	257, 456	
	Titanium dioxide	218, 279, 636	
	Tungsten and its compounds	338, 349, 351, 353	
	Water	199, 507, 522, 611, 742, 775, 778, 797, 839	
K	Ammonia synthesis catalysts	552	
	Beer		280
	Biological material	138, 142, 781	369, 516, 652, 811
	Cement		722, 748
	Chemical reagents	439	
	Fuel oil	751	
	Glass and ceramic material		417
	Fertilisers	34, 72, 84, 92, 129, 150, 267, 268, 269, 270, 271, 272, 273, 274, 301, 302, 303, 470, 504, 505, 526, 577, 635, 668, 699	
	Na–K alloys	208	
	Paints	117, 197, 416, 672, 700	
	Sodium and its compounds	264, 540	
	Soil	402, 428, 518, 519, 520, 527, 630, 638	

Appendix III Continued

Element analysed	Material analysed	Literature	
		Emission	Absorption
	Titanium dioxide	279	
	Uranium and its compounds	849	
	Vanadium contact mass	220	
	Water	138, 300, 405, 507, 522, 555, 611, 742, 776, 797	
Rb	Biological material	323	
	Glass	6, 806	
	Plants	323	740
	Silicates, ores, minerals	6, 37, 215, 250, 634, 654	45, 608
	Soil		740
	Water	145	111
	Whisky	605	
Cs	Bismuth and its compounds	800	
	Bi–U alloys	800	
	Glass	6, 806	
	Photocathodes	35	
	Silicates, ores, minerals	6, 37, 215, 249, 384, 469, 588, 634, 654	608
	Water	81, 145	
	Whisky	605	
Alkali metals	Aluminium	617	
	Ashes of various fuels	53, 407, 621, 685, 688	
	Alkali metal salts	376, 413, 583	
	Beer		280
	Biological material	68, 73, 82, 107, 122, 210, 242, 293, 342, 408, 530, 534, 610, 716, 772, 823	652
	Calcium and its salts	1, 627	
	Cement	31, 32, 46, 189, 235, 243, 462, 515, 563	722, 748
	Ceramic materials	29, 58, 148, 188, 235, 266, 311, 563, 631, 673	417
	Enamels	156	

Appendix III Continued

Element analysed	Material analysed	Literature	
		Emission	Absorption
	Fats	453	
	Glass	2, 35, 70, 96, 141, 248, 265, 341, 351, 352, 355, 357, 358, 360, 361, 362, 363, 473, 584, 604, 651, 653, 766, 801	417
	Graphite	40	
	Iron	461	
	Hydrogen peroxide	430	
	Petroleum	147	
	Plants	83, 551, 553	
	Photocathodes	35	
	Milk and cheese	343, 424	
	Silicates, ores, minerals	6, 37, 70, 99, 148, 149, 168, 181, 224, 228, 266, 332, 384, 386, 406, 429, 445, 467, 544, 556, 584, 588, 609, 665, 669, 744, 763, 803, 806	56, 608
	Soil	33, 113, 207, 240, 258, 291, 297, 549, 827	
	Tungsten and its compounds	349, 351	
	Uranyl nitrate	229	
	Water	81, 192, 193, 199, 507, 522, 554, 555, 595, 611, 687, 743, 797, 837	
	Wine	192, 193	
Cu	Air		625
	Aluminium and its alloys	786	54, 304
	Beer		280
	Biological material	789	60, 76, 372, 652, 704
	Alloys	172, 175, 176, 786	
	Brine	98	
	Fodder		347

Appendix III Continued

Element analysed	Material analysed	Literature	
		Emission	Absorption
	Gold	412	690, 734
	Iron		442, 692
	Cobalt concentrates	290	
	Glass	561	417
	Milk		532
	Minerals, ores	682	692, 733, 739
	Nickel solutions		694
	Plants	61	22, 90
	Potassium chloride		724
	Lubricants		115, 720
	Soil		22
	Steel	786	442
	Water		69, 111, 489, 578, 625
	Synthetic fertilisers		514
	Wine		841
Ag	Aluminium alloys		817
	Alloys		321
	Biological material		704
	Copper and its compounds	289	
	Gold	412	690, 734
	Lead concentrates		629
	Lubricants		115, 720
	Rocks		324
	Sulphide minerals		661
Au	Alloys		321
	Aqua regia		568
	Biological material		704
	Cyanide solutions		702
Ca	Antimony	539	
	Aluminium and silicon	459	

Appendix III Continued

Element analysed	Material analysed	Literature	
		Emission	Absorption
	Biological material	68, 107, 120, 122, 133, 186, 206, 210, 239, 286, 322, 329, 340, 342, 367, 388, 391, 408, 485, 486, 534, 601, 602, 612, 639, 650, 677, 678, 716, 754, 772, 783	652, 721, 759, 809, 812
	Beer		280
	Cellulose	569	
	Cement		722, 748
	Copper and its compounds	4	
	Ceramic material	29, 631	417
	Drugs	202, 483	
	Egg shells	498	
	Fertilisers	637	
	Foodstuffs	125	
	Fuel oil	531, 751	
	Glass	246, 653	7, 417
	Graphite	40	
	Hydrogen peroxide	430	
	Hafnium and Zirconium compounds	594	
	Carbonates	35	
	Lead	42	
	Loam	28	
	Fodder		347, 560
	Photocathodes	3, 351	
	Milk and cheese	343, 596	
	Nickel and its compounds	4	
	Petroleum	147	
	Plants	45, 83, 97, 143, 197, 305, 306, 346, 365, 404, 435, 450, 487, 553, 612, 680, 700, 732, 798, 807	160
	Rare earth oxides	364	

Appendix III Continued

Element analysed	Material analysed	Literature	
		Emission	Absorption
	Selenium	681	
	Silicates, ores, minerals	80, 355, 445, 451, 455, 467, 509, 533, 581	56, 337, 692
	Slag	2, 198, 396, 544, 727	327
	Soil	5, 113, 207, 240, 596, 612, 695, 756, 798, 827	161
	Uranium concentrates	257	
	Uranium and its compounds	729, 849	
	Tomatoes	479	
	Very pure acids	222	
	Water	137, 139, 199, 595, 612, 635, 741, 770, 777, 778, 797, 802	578
	Polyvinylchloride		537
	Tungsten and its compounds	349, 351, 353	
	Iron and its compounds	4, 326, 461	
	Zirconium and its compounds	543	
Mg	Aluminium alloys	395	54, 230, 472, 785
	Biological material	10, 68, 277, 671, 772, 789	131, 371, 418, 652, 721, 808, 810
	Cement	140	471, 722, 748
	Cider	295	
	Copper and its alloys	415	
	Fodder		347, 560
	Glass	653	7, 417
	Iron	461	50, 327, 692
	Limestone		337, 471, 657, 660
	Nickel and its compounds		27
	Plants	83, 382, 487, 671, 700	18, 90, 159, 223, 826
	Silicates, ores, minerals	80, 259, 495	56, 337, 660, 692

Appendix III Continued

Element analysed	Material analysed	Literature	
		Emission	Absorption
	Slag	2, 544, 653	327
	Soil	207, 382, 434, 671, 677 695, 827	161
	Synthetic fertilisers	514	
	Tomatoes	479	
	Uranium	513, 600	392
	Water	578	
	Wine	295	
	Whisky	606	
	Zinc alloys		529
Be	Bronze	493	
Sr	Biological material	380, 425, 677, 678, 783, 791	163, 165
	Carbonates	35, 458	
	Cement	190	
	Copper and its compounds	4	
	Coal ashes		51
	Electron-tube cathodes	3, 234, 351	
	Glass		7
	Iron and its compounds	4	
	Nickel and its compounds	4	
	Plants	436	
	Silicates, ores	275, 331, 443, 458, 581, 762	87
	Soil	436	165
	Water	136, 623, 708 742, 770, 778	111
Ba	Carbonates	35	
	Electron-tube cathodes	3, 334, 351	
	Glass		7, 417
	Barium sulphate	614	
	Lubricants	720	
	Iron and its compounds	4	
	Minerals		87
	Natural objects	240	

Appendix III Continued

Element analysed	Material analysed	Literature	
		Emission	Absorption
	Copper and its compounds	4	
	Nickel and its compounds	4	
	Uranium	599	
	Soil	683	
Alkaline earth elements	Electron-tube cathodes	234	
	Glass	71, 341	
	Soil	85	
Zn	Air		625
	Aluminium and its alloys		54, 55, 247, 304
	Biological material		131, 288, 381, 704, 814
	Fodder		347, 560
	Glass		417
	Graphite		546
	Bronze		201
	Iron		692
	Loamy materials		247
	Nickel solutions		694
	Ores		692, 739
	Plants		20, 101, 178
	Metallurgical material		310, 784
	Refractory metals		546
	Polyvinylchloride		537
	Synthetic fertilisers		514
	Silicates		110
	Wine		841
	Water		69, 111, 578, 381
Cd	Air		625
	Aluminium alloys		54
	Biological material		704, 705
	Iron and its ores		692

Appendix III Continued

Element analysed	Material analysed	Literature	
		Emission	Absorption
	Zinc and its ores		593
	Water		69, 625
	Steel		818
Hg	Biological material		813
	Fish, eggs		559
B	Borax	203	
	Fertilisers	84	
	B–Fe alloys	211	
	B–Zr alloys	211	
	Minerals, ores	171	
	Organic compounds	103, 835	
	Polishing baths	276	
Al	Alloys	428	496
	Bauxite		87
	Cement	422	
	Glass	358	
	Graphite, refractory metals		
	Kaolin	357	546
	Silicates	184	
	Trialkylaluminium		187
Ga	Alloys	521	
	Ores	74	
	Selenium	681	
In	Ores	74	
Tl	Biological material	728	705
	Minerals, ores	74, 838	
Cr	Chromium alloys	100	
	Aluminium alloys		54
	Biological material	158	256, 704, 705, 805
	Ferrites	799	
	Iron and iron ores		692
	Lubricants		115, 720
	Steel		41, 440
Se	Copper		723

Appendix III Continued

Element analysed	Material analysed	Literature	
		Emission	Absorption
Te	Copper		723
Mo	Biological material		704
Mn	Agrochemical material		19, 514
	Air		664
	Aluminium alloys	176	54
	Biological material		704
	Cement	191	748
	Ceramic material		417
	Ferrites	799	
	Fodder		347
	Gasoline	713	
	Glass	653	417
	Iron		692
	Minerals, ores, silicates	196	56, 692
	Slag	544	
	Steel		52
	Plants	61, 432	101
	Water		111, 664
Fe	Agrochemical material		19, 514
	Air		664
	Alloys	174	54
	Beer		280
	Biological material		652, 704
	Cement		722
	Cobalt concentrates	290	
	Fodder		347
	Glass and ceramic material		417
	Gold		690, 734
	Ferrites	799	
	Lubricants		115, 720
	Nickel solutions		694
	Ores, silicates, minerals	173	56, 337
	Plants	61, 159	
	Water		69, 111, 578, 664

Appendix III Continued

Element analysed	Material analysed	Literature	
		Emission	Absorption
Co	Agrochemical material		21
	Air		664
	Biological material		704, 705
	Cobalt concentrates	290	
	Glass and ceramic material		417
	Iron ores and iron		692
	Nickel		517
	Steel		517
	Water		108, 111, 664
Ni	Agrochemical material		21
	Air		664
	Aluminium alloys	176	54
	Biological material		111, 704, 814
	Glass and ceramics		417, 664
	Iron ores, iron and steel		441, 692
	Oil		758
	Polishing baths	276	
	Water		69, 108, 111, 578, 664
Pd	Alloys		321
	Gold		734
Rh	Alloys		321
Pt	Alloys		321
P	Ores	195	
	Organic substances	95, 492	
	Phosphates	62, 628, 637	
S	Biological material	696, 697	
	Degreasing powders	153	
	Water	696	
Cl	Organic substances	502	
Y	Rare earth oxides	558	
Gd	Rare earth oxides	558	
La	Glass	400	
	Ores	400	
	Uranium	600	

Appendix III Continued

Element analysed	Material analysed	Literature	
		Emission	Absorption
Si	Bauxite		87
Sn	Alloys		121
	Hydrogen peroxide		9
Pb	Air		750
	Alloys	130	157, 231
	Biological material		704, 705, 812
	Fish meal		735
	Gasoline	313, 713	642, 816
	Gold		690
	Iron		692
	Lubricants		115, 720
	Glass and ceramic material		417
	Nickel solutions		694
	Ores		692, 739
	Polyvinylchloride		537
	Potassium chloride		724
	Steel		157, 231
	Water		69
	Wine		841
Ti	Alloys, steel, iron		87, 345
	Bauxite		87
Zr	Thorium oxides		764
Sb	Alloys		535
Bi	Ores, iron		692
V	Oils		87

REFERENCES

1. Aasness H.: Norsk. farm. Selsk. **19**, 53 (1957)
2. Abresch K., Dobner W.: Arch. Eisenhüttenw. **29**, 25 (1958)
3. Adám J., Ettre K., Gergely G., Varadi P. F.: Mag. Kém. Foly. **62**, 223 (1956)
4. Adám J., Ettre K.: Mag. Kém. Foly **63**, 206 (1957)
5. Adams F., Rouse R. D.: Soil Sci. **83**, 305 (1957)
6. Adams P. B.: Anal. Chem. **33**, 1602 (1961)
7. Adams P. B., Passmore W. O.: Anal. Chem. **38**, 630 (1966)
8. Adams W. G., Day R. E.: Proc. Roy. Soc. (London) A **25**, 113 (1877)
9. Agazzi E. J.: Anal. Chem. **37**, 364 (1965)
10. Alcock N., MacIntyre I., Radde I.: J. Clin. Pathol. **13**, 506 (1960)
11. Alkemade C. T. J.: Thesis, University of Utrecht 1954
12. Alkemade C. T. J.: Colloquium Spectroscopicum Internationale VI., Pergamon Press, London, 1957
13. Alkemade C. T. J., Jeuken M. E. J.: Z. anal. Chem. **158**, 401 (1957)
14. Alkemade C. T. J., Voorhuis M. H.: Z. anal. Chem. **163**, 91 (1958)
15. Alkemade C. T. J.: VIII. Colloquium Spectroscopicum Internationale (Luzerne 1959), Sauerländer, Aarau, Switzerland 1960
16. Alkemade C. T. J.: Proc. Xth Coll. Spectr. Int., Maryland 1962, p. 143
17. Alkemade C. T. J.: Anal. Chem. **38**, 1252 (1966)
18. Allan J. E.: Analyst **83**, 466 (1958)
19. Allan J. E.: Spectrochim. Acta **15**, 800 (1959)
20. Allan J. E.: Nature **187**, 1110 (1960)
21. Allan J. E.: Analyst **86**, 530 (1961)
22. Allan J. E.: Spectrochim. Acta **17**, 459 (1961)
23. Allan J. E.: Spectrochim. Acta **17**, 467 (1961)
24. Allan J. E.: Spectrochim. Acta **18**, 259 (1962)
25. Allan J. E.: Spectrochim. Acta **18**, 605 (1962)
26. Amos M. D., Willis J. B.: Spectrochim. Acta **22**, 1325 (1966)
27. Andrew T. R., Nichols P. N. R.: Analyst **87**, 25 (1962)
28. Anon.: Brit. Clayworker **64**, 160 (1955)
29. Anon.: Trans. Brit. Ceram. Soc. **54**, No. 6 (1955)
30. Arthur J. A.: Progress in Coal Science 365 (1949)
31. ASTM: Methods for Emission Spectrochemical Analysis, ASTM publ., Philadelphia 1953

32. ASTM Standards, Part III, 157 (1952)
33. ATTOE O. J., TRUOG E.: Soil Sci. Soc. Amer., Proc. **11**, 221 (1946)
34. AUSTIN H. C. jun. and Ass.: J. Assoc. Offic. Agr. Chemists **36**, 885 (1953)
35. AVERBUKH M. A.: Zavodsk. Lab. **27**, 358 (1961)
36. AVNI R., ALKEMADE C. T. J.: Microchim. Acta 460 (1960)
37. AXELROD J. M., ADLER I.: Anal. Chem. **29**, 1280 (1957)
38. BAKER C. A., CARTON F. W. J.: U.K.A.E.A. Report 3490
39. BAKER M. R., VALLEE B. L.: Anal. Chem. **31**, 2036 (1959)
40. BARANSKA H.: Chem. anal. (Warsaw) **2**, 229 (1957)
41. BARNES L. jr.: Anal. Chem. **38**, 1083 (1966)
42. BARRINGER R. E.: U.S. Atomic Energy Comm. Rep. Y – 1295 (1960)
43. BARROW R. G. J.: J. Med. Lab. Technol. **17**, 236 (1961)
44. BAUSERMAN H. M., CERNEY R. R.: Anal. Chem. **25**, 1821 (1953)
45. BAZHOV A. S., LAZAREV J. A., KOKA P. A.: Zavodsk. Lab. **33**, 173 (1967)
46. BECKER F.: Zement–Kalk–Gibs **4**, 93 (1951)
47. Beckman Instruments, Inc., Application Data DU-12-B
48. Beckman Instruments, Inc., Bulletin 3063 (1962)
49. BECKMANN E., WAENTIG P.: Z. physik. Chem. **68**, 385 (1910)
50. BELCHER C. B., BRAY H. M.: Anal. Chim. Acta **26**, 322 (1962)
51. BELCHER C. B., BROOKS K. A.: Anal. Chim. Acta **29**, 202 (1963)
52. BELCHER C. B., KINSON K.: Anal. Chim. Acta **30**, 483 (1964)
53. BELCHER R. and Assoc.: Anal. Chim. Acta **11**, 120 (1954)
54. BELL C. F.: Atomic Absorp. Newsletter **5**, 73 (1966)
55. BELL G. F. Atomic Absorp. Newsletter **6**, 118 (1967)
56. BELT CH. B. jun.: Anal. Chem. **39**, 676 (1967)
57. BENJAMINSEN J., JENSEN J.: Tiddskr. Planteavl. **60**, 43 (1956)
58. BENNETT H., EARDLEY R. P., HAWLEY W. G., THWAITES J.: Trans. Brit. Ceram. Soc. **61**, 433 (1962)
59. BERÁNEK J.: Radioisotopy No 12–14, 123
60. BERMAN E.: Atomic Absorp. Newsletter **4**, 296 (1965)
61. BERNKING A. D., SCHRENK W. G.: J. Agric. Food Chem. **5**, 742 (1957)
62. BERNHARDT D., HERRMANN R.: Aertl. Wochschr. **10**, 61 (1955)
63. BERNSTEIN R. E.: S. African J. Med. Sci. **14**, 163 (1949)
64. BERNSTEIN R. E.: Biochim. et Biophys. Acta **9**, 567 (1952)
65. BERNSTEIN R. E.: Nature **165**, 549 (1950)
66. BERRY J. W., CHAPPELL D. G., BARNES R. B.: Ind. Eng. Chem. Anal. Ed. **18**, 19 (1946)
67. BEUKELMAN T. E., LORD S. S. jr.: Appl. Spectroscopy **14**, 12 (1960)
68. BIANCHI E.: Boll. soc. ital. biol. sperim. **32**, 1129 (1956)
69. BIECHLER D. G.: Anal. Chem. **37**, 1054 (1965)
70. BIFFEN F. M.: Anal. Chem. **22**, 1014 (1950)
71. BILLINGS D.: Glass Ind. **36**, 255, 280 (1956)
72. BLACKWELL A. T., YEAGER C. L., KRAUS M.: J. Assoc. Offic. Agr. Chemists **36**, 898 (1953)

73. BLUM A. S.: Nucleonics **14**, 64 (1956)
74. BODE H., FABIAN H.: Z. anal. Chem. **170**, 387 (1959)
75. BOLAND F. E.: Jour. A.O.A.C. **49**, 617 (1966)
76. BOLING E. A.: Anal. Chem. **37**, 482, (1965)
77. BOROVIK-ROMANOVA T. F.: Tr. Biogeokhim. Lab. A.N. USSR **8**, 157 (1946)
78. BOROVIK-ROMANOVA T. F.: Zh. analit. Khim. **16**, 664 (1961)
79. BORZOV V. P., PLYUSH G. V., YANKOVICH N. G.: Zavodsk. Lab. **32**, 1467 (1966)
80. BOSCH H.: Tonind. Ztg. Keram. Rundschau **81**, 7 (1957)
81. BOSSUET R.: Compt. Rendus **198**, 1094 (1935)
82. BOTT P. A.: Anal. Biochem. **1**, 17 (1960)
83. BOVAY E.: Mitt. Gebiete Lebensm. u. Hyg. **46**, 540 (1955)
84. BOVAY E., COSSY A.: Mitt. Gebiete Lebensm. u. Hyg. **48**, 59 (1957)
85. BOWER C. A., REITMAIER R. F.: Soil. Sci. **73**, 251 (1952)
86. BOWMAN J. A.: Anal. Chim. Acta **37**, 465 (1967)
87. BOWMAN J. A., WILLIS J. B.: Anal. Chem. **39**, 1210 (1967)
88. BOX G. F., WALSH A.: Spectrochim. Acta **16**, 225 (1960)
89. BRABSON J. A., WILHIDE W. D.: Anal. Chem. **26**, 1060 (1954)
90. BRADFIELD E. G., SPINCER D.: J. Sci. Fd. Agric. **16**, 33 (1965)
91. BRAICOVICH L., LANDI M. F.: Spectrochim. Acta **11**, 51 (1957)
92. BREALEY L.: Analyst **76**, 340 (1951)
93. BREALEY L., GARRAT D. C., PROCTOR K, A.: J. Pharm. and Pharmacol. **4**, 717 (1952)
94. BRHÁČEK L., GOLONKA A., JANÁČEK J.: Hutnické listy **13**, 719 (1958)
95. BRITE D. W.: Anal. Chem. **27**, 1815 (1955)
96. BRODERICK E. J., ZACK P. G.: Anal. Chem. **23**, 1455 (1951)
97. BROGAN J. C.: J. Sci. Food Agric. **11**, 446 (1960)
98. BROOKS R. R., PRESLEY B. J., KAPLAN I. R.: Anal. Chim. Acta **38**, 321 (1967)
99. BRUMBAUGH R. J., FANUS W. E.: Anal. Chem. **26**, 464 (1954)
100. BRYAN H. A., DEAN J. A.: Anal. Chem. **29**, 1289 (1957)
101. BUCHANAN J. R., MARAOKA T. T.: Atomic Absorp. Newsletter 1964 No 24, p. 1
102. BÜCKERT H., RAFAELE I.: Internationales Jahrbuch Chemische Industrie, 1966, Vogt—Schild, Solothurn.
103. BUELL B. E.: Anal. Chem. **30**, 1514 (1958)
104. BUELL B. E.: Anal. Chem. **34**, 635 (1962)
105. BULEWICZ E. M., PHILLIPS L. F., SUGDEN T. M.: Trans. Faraday Soc. **57**, 921 (1961)
106. BULEWICZ E. M., SUGDEN T. M.: Trans. Faraday Soc. **54**, 830 (1958)
107. BÜNTE H.: Aerztl. Lab. **5**, 345 (1959)
108. BURRELL D. C.: Atomic Absorp. Newsletter **4**, 309 (1965)
109. BURRELL D. C.: Atomic Absorp. Newsletter **4**, 328 (1965)
110. BURRELL D. C.: Norsk. geol. Tidsskr. **45**, 21 (1965)
111. BURRELL D. C.: Anal. Chim. Acta **38**, 447 (1967)
112. BURRIEL-MARTI F.: Experientia (1956), Suppl. No 5, p. 71

113. BURRIEL-MARTI F., RAMIREZ-MUNOZ J., VENITO-POTOUS A.: Anal. Real. Soc Espan. Física Quim. Ser B, **53**, 521 (1957)
114. BURRIEL-MARTI F., REXACH-M., de LIZARDUY M. L., RAMIREZ-MUNOZ J.: Rev. Univ. ind. Santander **1**, 37 (1959)
115. BURROWS J. A., HEERDT J. C., WILLIS J. B.: Anal. Chem. **37**, 579 (1965)
116. BURYAK N. I., MAKAROVA E. J.: Sb. Vanadievie katalizatory, Goskhimizdat, Moscow 1963
117. BUSSMANN A.: Z. Pflanzenernähr. Düng. u. Bodenk. **77**, 212 (1957)
118. BUTLER L. R. P.: S. Afr. Ind. Chemist **15**, (1961)
119. BUTLER L. R. P.: J. S. Afr. Inst. Min. Metallurgy **62**, 780 (1962)
120. BUTTERWORTH E. C.: J. Clin. Pathol. **10**, 379 (1957)
121. CAPACHO-DELGADO L., MANNING D. C.: Atomic Absorp. Newsletter **4**, 317 (1965)
122. CARDÚS D., LLAURADO J. G.: Med. Clin. (Barcelona) **24**, 193 (1955)
123. CARNES W. J., DEAN J. A.: Anal. Chem. **33**, 1961 (1961)
124. CARNES W. J., DEAN J. A.: Analyst **87**, 748 (1962)
125. CAROLAN R.: Int. Sugar J. **61**, 103 (1959)
126. CARTWRIGHT J. S., SEBENS C., MANNING D. C.: Atomic Absorp. Newsletter **5**, 84 (1966)
127. CASTLEMAN R. A.: J. Research Natl. Bur. Standards **6**, 369 (1931)
128. CATON R. D., BREMNER R. W.: Anal. Chem. **26**, 805 (1954)
129. CAVELL A. J.: Analyst **77**, 537 (1952)
130. CHAKRABARTI C. L., MAGEE R. J., WILSON C. L.: Talanta **10**, 57 (1963)
131. CHANG T. L., GOVER T. A., HARRISON W. W.: Anal. Chim. Acta **34**, 17 (1966)
132. CHATONIER D.: Bull. Soc. Pharm. Bordeaux **94**, No. special, 29 (1955)
133. CHEN P. S. jun., TORIBARA T. Y.: Anal. Chem. **25**, 1642 (1955)
134. CHOINIERE L.: Canad. J. Agric. Sci. **36**, 203 (1956)
135. CHOLAK J., HUBBARD D. M.: Ind. Eng. Chem., Anal. Ed. **16**, 728 (1944)
136. CHOW T. J., THOMPSON T. G.: Anal. Chem. **27**, 18 (1955)
137. CHOW T. J., THOMPSON T. G.: Anal. Chem. **27**, 910 (1955)
138. CHOW T. J.: Anal. Chim. Acta **31**, 58 (1964)
139. CHRIST W.: Chem. Technik **8**, 280 (1956)
140. CLOSE P., SMITH W. E., WATSON M. T. jun.: Anal. Chem. **25**, 1022 (1953)
141. CLOSE P., WATSON M. T. jun.: J. Am. Ceram. Soc. **37**, 235 (1954)
142. COBO DE CELIS M. E., MARTIN A., PACHEO B.: Rev. Assoc. Bioquim. Argentina **25**, 95 (1960)
143. COOLEY M. L.: Cereal. Chem. **30**, 39 (1953)
144. COLLIER H. E.: Thesis, Lehigh Univ. 1955
145. COLLINS A. G.: Anal. Chem. **35**, 1258 (1963)
146. CONNORS J. J.: J. Am. Water Works Assoc. **42**, 39 (1950)
147. CONRAD A. L., JOHNSON W. C.: Anal. Chem. **22**, 1530 (1950)
148. CORADUSCI P.: Ceramica (Milano) N. S. **10**, 37 (1955)
149. COREY R. B., JACKSON M. L.: Anal. Chem. **25**, 624 (1953)
150. CROOKS R. C.: J. Assoc. Offic. Agr. Chemists **36**, 891 (1953)

151. ČSN 65 1290 Kyselina wolframová technická (Czechoslovak Standard technical tungstic acid)
152. Cullum D. C., Thomas D. B.: Analyst **84**, 113 (1959)
153. Cullum D. C., Thomas D. B.: Analyst **85**, 688 (1960)
154. Cunningham E.: Proc. Roy. Soc., Ser. A, **83**, 357 (1910)
155. Curtis G. W., Knauer H. E., Hunter L. E.: Symposium on Flame Photometry, Sp. Techn. Publ. ASTM **116**, 67 (1952)
156. Czajka Z., Chem. anal. (Warsaw) **7**, 355 (1962)
157. Dagnal R. M., West T. S., Young P.: Anal. Chem. **38**, 358 (1966)
158. Daly J. R., Anstall H. B.: Clin. Chim. Acta **9**, 576 (1964)
159. David D. J.: Analyst **83**, 655 (1958)
160. David D. J.: Analyst **84**, 536 (1959)
161. David D. J.: Analyst **85**, 495 (1960)
162. David D. J.: Analyst **85**, 779 (1960)
163. David D. J.: Nature **187**, 1109 (1960)
164. David D. J.: Analyst **86**, 730 (1961)
165. David D. J.: Analyst **87**, 576 (1962)
166. David D. J.: Rev. Univ. Ind. de Santander **4**, 207 (1962)
167. David D. J.: Analyst **89**, 747 (1964)
168. Davis C. E. S.: Soc. Chem. Ind. Victoria (Proc.) **51**, 1032 (1952)
169. Davis H. M., Fox G. P., Webb R. J., Wildy P. C.: U.K.A.E.A., A.E.R.E. Rep. R. 2659, 1961
170. Dean J. A., Thompson C.: Anal. Chem. **26**, 1855 (1954)
171. Dean J. A., Thompson C.: Anal. Chem. **27**, 42 (1955)
172. Dean J. A.: Anal. Chem. **27**, 1224 (1955)
173. Dean J. A., Burger J. C. jun.: Anal. Chem. **27**, 1052 (1955)
174. Dean J. A., Lady J. H.: Anal. Chem. **27**, 1533 (1955)
175. Dean J. A., Lady J. H.: Anal. Chem. **28**, 1885 (1956)
176. Dean J. A., Cain C. jun.: Anal. Chem. **29**, 530 (1957)
177. Dean J. A.: Flame Photometry, McGraw-Hill Book Comp., Inc., New York, Toronto, London 1960
178. Dean J. A., Stubblefield Ch. B.: Anal. Chem. **33**, 382 (1961)
179. Dean J. A., Carnes W. J.: Anal. Chem. **34**, 192 (1962)
180. Dean J. A., Carnes W. J.: Analyst **87**, 743 (1962)
181. Debras J., Voinovitch I. A.: Bull. Soc. Franc. Céram. **38**, 77 (1958)
182. Debras J., Voinovitch I. A.: Chem. anal. (Warsaw) **3**, 303 (1958)
183. Debras J., Voinovitch I. A.: Compt. rend. **248**, 77 (1959)
184. Debras-Guédon J., Voinovitch I. A.: Compt. rend. **249**, (1959)
185. Debras J., Voinovitch I. A.: Rev. Univ. Mines 9 ser. T XV, 408 (1959)
186. Dehority B. A., J. Dairy Sci. **42**, 872 (1959)
187. Deily J. R.: Atomic Absorp. Newsletter **5**, 119 (1966)
188. Diamond J. J., Bean L.: Symposium on Flame Photometry, Sp. Techn. Publ. ASTM **116**, 28 (1952)

189. Diamond J. J., Bean L.: Anal. Chem. **25,** 1825 (1953)
190. Diamond J. J.: Anal. Chem. **27,** 913 (1955)
191. Diamond J. J.: Anal. Chem. **28,** 328 (1956)
192. Diemair W., Gundermann C.: Z. Lebensmittel-Unters. u. Forsch. **111,** 120 (1959)
193. Diemair W., Gundermann C.: Z. Lebensmittel-Unters. u. Forsch. **109,** 469 (1959)
194. Dinnin J. I.: Anal. Chem. **32,** 1475 (1960)
195. Dippel W. A., Bricker C. E., Furman N. H.: Anal. Chem. **26,** 553 (1954)
196. Dippel W. A., Bricker C. E.: Anal. Chem. **27,** 1484 (1955)
197. Dobbins J. T. jun.: Jour. A.O.A.C. **49,** 521 (1966)
198. Dobrowolski J., Wyszynski N.: Rocznik Pomorskiej Akad. Med. Warsaw **3,** 403 (1957)
199. Dobřemyslová I., Dvořák J., Novobilský V.: Chem. průmysl **12,** 287 (1962)
200. Doerffel K., Geyer R., Müller G.: Chem. anal. (Warsaw) **7,** 229 (1962)
201. Doerffel K., Nitzsche U.: Wiss. Z. tech. Hochsch. Chem. Leuna Mersb. **7,** 9 (1965)
202. Dorche J., Costet C.: Ann. pharm. franc. **14,** 669 (1956)
203. Dorche J. and Assoc.: Bull. trav. soc. pharm. Lyon **1,** 93 (1958)
204. Dowling F. B., Chakrabarti C. L., Lyles G. R.: Anal. Chim. Acta **28,** 392 (1963)
205. Dragoun Z., Šmirous K.: Semiconductors (in Czech) SNTL Praha, 1959
206. Dreisbach R. H.: Rev. Univ. Ind. de Santander **3,** 7 (1961)
207. Dubrovin K. P.: Thesis, Univ. of Wisconsin, Madison, U.S.A. 1956
208. Dubrovina T. P., Berdicheskii N. I.: Zavodsk. Lab. **32,** 166 (1966)
209. Dunken H., Pforr G. and Assoc.: Z. Chem. **3,** 196 (1963)
210. Duthie I. F., MacDonald I.: Lab. Practice **9,** 705 (1960)
211. Dutina D.: Anal. Chem. **30,** 2006 (1958)
212. Dvořák J., Řezáč Z.: Chem. listy **53,** 588 (1959)
213. Dvořák J., Řezáč Z.: Chem. listy **53,** 1113 (1959)
214. Dvořák J.: Chem. listy **54,** 28 (1960)
215. Dvořák J.: Chem. průmysl **11,** 122 (1961)
216. Dvořák J., Novobilský V.: Acta Chim. Acad. Sci. Hung. **30,** 365 (1962)
217. Dvořák J., Řezáč Z.: Acta geol. geograph. Univ. Comenianae No 6, 499 (1962)
218. Dvořák J., Truplová E.: Coll. Czech. Chem. Commun. **28,** 1609 (1963)
219. Dvořák J., Rubeška I.: Chem. listy **57,** 561 (1963)
220. Dvořák J.: not published
221. Dvořák J., Krejcarová M., Kučerová J., Žižka B.: Coll. Czech. Chem. Commun. **31,** 1881 (1966)
222. Dvořák J., Míšková I., Mareček J., Ditz J.: Chem. průmysl **40,** 308 (1965)
223. Dvořák J.: Chem. listy **61,** 390 (1967)
224. Easton A. J., Lovering J. F.: Anal. Chim. Acta **30,** 543 (1964)
225. Eckhard S., Püschel A.: Z. anal. Chem. **172,** 334 (1960)

226. Eckschlager K.: Error of Chemical Analyses, p. 96 (in Czech) 1961
227. Edgcombe L. J., Hewett D. R.: Analyst **79,** 755 (1954)
228. Ellestad R. B., Horstman E. L.: Anal. Chem. **27,** 1229 (1955)
229. Ellis W. G., Brown E. A.: U.S. At. Energy Comm. TID 7568, Pt 1, 205 (1958)
230. Elwell W. T., Gidley J. A. F.: Atomic absorption spectrophotometry, p. 26, Pergamon Press, London 1961
231. Elwell W. T., Gidley J. A. F.: Anal. Chim. Acta **24,** 71 (1961)
232. Erdey L., Svehla G.: Z. anal. Chem. **154,** 406 (1957)
233. Eshelman H. C., Dean J. A.: Anal. Chem. **33,** 1339 (1961)
234. Ettre K., Adám J.: Z. anal. Chem. **155,** 105 (1957)
235. Eubank W. R., Bogue R. H.: J. Research Natl. Bur. Standards **43,** 173 (1949)
236. Evans Electroselenium Ltd., Harlow, Essex, Publ. No F 117 and 132
237. Evans Electroselenium Ltd., Reference: D1
238. Evans Electroselenium Ltd., Reference: D2
239. Evans Electroselenium Ltd., Reference: D3
240. Evans Electroselenium Ltd., Reference: D4
241. Evans Electroselenium Ltd., Reference: D5
242. Evans Electroselenium Ltd., Reference: D6
243. Evans Electroselenium Ltd., Reference: D7 and D7/1
244. Evans Electroselenium Ltd., Reference: D8
245. Evans Electroselenium Ltd., Reference: D12
246. Evans Electroselenium Ltd., Reference: D15
247. Ezell J. B. jun.: Atomic Absorp. Newsletter **5,** 122 (1966)
248. Fabricand B. P. et al.: Geochim. Cosmochim. Acta 1023 (1962)
249. Fabrikova E. A.,: Zh. analit. Khim. **14,** 41 (1959)
250. Fabrikova E. A.: Zh. analit. Khim. **15,** 427 (1960)
251. Fabrikova E. A.: Zh. analit. Khim. **16,** 22 (1961)
252. Fabrikova E. A., Isaeva A. G.: Zh. analit. Khim. **18,** 329 (1963)
253. Fassel V. A., Mossotti V. G.: Anal. Chem. **35,** 252 (1963)
254. Fassel V. A., Myers R. B., Kniseley R. N.: Spectrochim. Acta **19,** 1187 (1963)
255. Fassel V. A., Golightly D. W.: Anal. Chem. **39,** 466 (1967)
256. Feldman F. J., Knoblock E. C., Purdy W. C.: Anal. Chim. Acta **38,** 489 (1967)
257. Fergason L. A.: U.S. At. Energy Comm. TID-7567, Pt 1, 197 (1958)
258. Fields M., King P. J. T., Richardson J. P., Swindale L. D.: Soil. Sci. **72,** 219 (1951)
259. Fink A.: Mikrochim. Acta **314** (1955)
260. Finkelstein W. P., Jansen A. V.: S. Afr. Ind. Chemist **15,** 106 (1961)
261. Fischer J., Doiwa A.: Spectrochim. Acta **11,** 28 (1957)
262. Fischer J., Kropp R.: Glastechn. Ber. **33,** 380 (1960)
263. Fischer J., Kropp R.: Z. anal. Chem. **179,** 1 (1961)
264. Fischer A. M., Finkelstein A. I.: Zavodsk. Lab. **23,** 788 (1957)
265. Fletscher W. W.: Transact. Soc. Glass. Technol. **43,** 86 (1959)
266. Ford C. L.: Anal. Chem. **26,** 1578 (1954)

267. Ford O. W.: J. Assoc. offic. Agr. Chemists **33,** 268 (1950)
268. Ford O. W.: J. Assoc. Offic. Agr. Chemists **34,** 660 (1951)
269. Ford O. W.: J. Assoc. Offic. Agr. Chemists **35,** 674 (1952)
270. Ford O. W.: J. Assoc. Offic. Agr. Chemists **36,** 649 (1953)
271. Ford O. W.: J. Assoc. Offic. Agr. Chemists **37,** 363 (1954)
272. Ford O. W.: J. Assoc. Offic. Agr. Chemists **38,** 445 (1955)
273. Ford O. W.: J. Assoc. Offic. Agr. Chemists **39,** 598 (1956)
274. Ford O. W.: J. Assoc. Offic. Agr. Chemists **40,** 722 (1957)
275. Fornaseri M., Grandi L.: Geochim. et Cosmoch. Acta **19,** 218 (1960)
276. Fornwalt D. E.: Anal. Chim. Acta **17,** 597 (1957)
277. Van Fossan D. D., Baird E. E.: Am. J. Clin. Pathol. **31,** 368 (1959)
278. Foster W. H., Hume D. N.: Anal. Chem. **31,** 2028 (1959)
279. Fratkin Z. G., Moshkovich G. N., Filippova Z. A.: Zavodsk. lab. **31,** 1090 (1965)
280. Frey S. W.: Atomic Absorp. Newsletter **3,** 127 (1964)
281. Fukushima S., Shigemoto M., Kato I., Otozai K.: Mikrochim. Acta 298 (1955)
282. Fukushima S., Iwata S., Kume S., Shigemoto M.: Japan Analyst **5,** 704 (1956)
283. Fukushima S., Takahashi K., Terasaka S., Otozai K.: Mikrochim. Acta 183 (1957)
284. Fukushima S., Yukawa K., Shigemoto M., Otozai K.: Mikrochim. Acta 553 (1958)
285. Fukushima S.: Mikrochim. Acta 596 (1959)
286. Fukushima S.: Mikrochim. Acta 332 (1960)
287. Fuwa K., Thiers R. E., Vallee B. L., Baker M. R.: Anal. Chem. **31,** 2039 (1959)
288. Fuwa K., Pulido P., McKay R., Vallee B. L.: Anal. Chem. **36,** 2407 (1964)
289. Galloway N. McN.: Analyst **83,** 373 (1958)
290. Galloway N. McN.: Analyst **84,** 505 (1959)
291. Gammon N.: Soil. Sci. **71,** 211 (1951)
292. Gamsjäger H., Ornig H., Schwarz-Bergkampf E.: Mikrochim. Acta 607 (1957)
293. Garciá-Llauradó J.: Med. Clin. (Barcelona) **21,** 250 (1953)
294. Garman V. F., U.S. Patent **2**, 836097 (1958)
295. Gärtel W.: Weinberg u. Keller **4**, 534 (1957)
296. Gatehouse B. M., Willis J. B.: Spectroch. Acta **17,** 710 (1961)
297. Gattorta G., ServelloV.: Ann. Staz. Chim. Agr. Roma **3,** 15 (1958)
298. Gaydon A. G., Wolfhard H, G.: Flames, their Structure, Radiation and Temperature. Chapman and Hall, London (1953)
299. Gaydon A. G.: The Spectroscopy of Flames. Chapman and Hall, London (1957)
300. Gazzi V.: Chim. ind. **36**, 249 (1954)
301. Gehrke C. W., Affsprung H. W., Wood E. L.: J. Agr. Food Chem. **3,** 48 (1955)
302. Gehrke C. W., Wood E. L.: Missouri Univ. Agr. Expt. Sta., Research Bull., No. 635 (1957)

303. GELBI P.: Ann. sper. agrar. **6,** 359 (1952)
304. GERBATSCH R.: Emissionspektroskopie, Akademie-Verlag—Berlin, **181,** (1964)
305. GETTHANDT G.: Z. Pflanzenernähr., Düng. u. Bodenk. **74,** 135 (1956)
306. GETTHANDT G.: Z. Pflanzenernähr., Düng. u. Bodenk. **78,** 187 (1957)
307. GETTHANDT G.: Landwirtsch. Forsch. **11,** 93 (1958)
308. GIANNINI G. M.: Sci. American 80 (1957)
309. GIBSON J. H., GROSSMAN W. E. L., COOKE W. D.: Anal. Chem. **35,** 266 (1963)
310. GIDLEY J. A. F., JONES J. T.: Analyst **85,** 249 (1960)
311. GIESEN K., KAMPA P.: Tonind.-Ztg. keram. Rdsch. **77,** 383 (1953)
312. GILBERT P. T., HAWES R. C., BECKMAN A. O.: Anal. Chem. **22,** 772 (1950)
313. GILBERT P. T.: Anal. Chem. **23,** 1053 (1951)
314. GILBERT P. T.: Symposium on Flame Photometry, Sp. Techn. Publ. ASTM **116,** 77 (1952)
315. GILBERT P. T.: Beckman Bulletin 753-A Flame Spectra of the Elements
316. GILBERT P. T.: Analyser **2,** 3 (1961)
317. GILBERT P. T.: Anal. Chem. **34,** 1025 (1962)
318. GILBERT P. T.: Proceedings of X. Colloquium Spectroscopicum Internationale (1962)
319. GILBERT P. T.: Anal. Chem. **38,** 1920 (1966)
320. GILLILAND J.: Symposium on Flame Photometry ASTM Techn. Publ. **116,** 33 (1951)
321. GINSBURG V. L., LIVSHITS D. M., SATARINA G. I.: Zh. analit. khim. **19,** 1089 (1964)
322. GLASMACHER H.: Arzneimittel-Forsch. **9,** 671 (1959)
323. GLENDENING B. L., PARISH D. B., SCHRENK W. G.: Anal. Chem. **27,** 1554 (1955)
324. GOMEZ COEDO A., JIMENEZ SECO J. L.: Revta Metal. **1,** 158 (1965)
325. GÖRLICH P.: Die lichtelektrischen Zellen. Akademische Verlagsgesellschaft, Leipzig 1951
326. GOTO H., IKEDA S., KIMURA J.: J. Japan Inst. Metals, Sendai **22,** 185 (1958)
327. GOTO H., IKEDA S., ATUYA I.: Japan Analyst **13,** 111 (1964)
328. GOUY C. L.: Ann. chim. et phys. **18,** 5 (1879)
329. GREKH J. F., BOGNIBOV E. A.: Lab. Delo **8,** 15 (1962)
330. GRIMALDI F. S.: GEOL. Surv. Research Prof. Paper 400-B (1960)
331. GUNDLACH H.: Z. anal. Chem. **171,** 9 (1959)
332. GURVICH I. G., KHANAEV J. I.: Izv. Acad. Sci. USSR Geol. Ser. **22,** 101 (1957)
333. GURVICH I. G., KHANAEV J. I.: Izv. Acad. Sci. USSR Geol. Ser. **22,** 104 (1957)
334. HALLS D. J., TOWNSHEND: Anal. Chim. Acta **36,** 278 (1966)
335. HALPERIN A., SAMBURSKY S.: J. Opt. Soc. Am. **42,** 472 (1952)
336. HAMAKER H. C., BEEZHOLD W. F.: Physica **1,** 119 (1933)
337. HAMEAU G.: Afinidad **22,** 176 (1965)
338. HAMERNÍK R.: Hutn. listy **14,** 142 (1959)
339. HARRISON G. E.: Photoelect. Spectr. Gr. Bull. **327** (1959)
340. HARRISON G. E., SUTTON A.: Nature **197,** 809 (1963)

341. HARTLEY F.: Proc. Internat. Comm. Glass **2,** 45 (1955)
342. HÄUSSLER, HAJDÚ P.: Arch. Pharmazie **292,** 73 (1959)
343. HAVE A. J. van der, MULDER H.: Neth. Milk Dairy J. **11,** 128 (1957)
344. HAVELKA B., KEPRT E.: Spektrální analýza I. NČSAV, Praha 1957
345. HEADRIDGE J. B., HUBBARD D. P.: Anal. Chim. Acta **37,** 151 (1967)
346. HECKMAN M.: J. Assoc. Offic. Agr. Chemists **43,** 337 (1960)
347. HECKMAN M.: Journ. A.O.A.C. **50,** 45 (1967)
348. HEENEY H. B., WARD G. M., WILLSON A. F.: Analyst **87,** 49 (1962)
349. HEGEDÜS A. J., DVORSZKY M.: Ber. Forschungsinst. Nachrichtentechn. Ind. Nr **30,** Budapest 1953
350. HEGEDÜS A. J., FUKKER F. K., DVORSZKY M.: Mag. Kém. Foly. **59,** 334 (1954)
351. HEGEDÜS A. J.: Dissert. Ung. Akad. Wiss. Abt. Chem. Wiss., Budapest 1958
352. HEGEDÜS A. J., DVORSZKY M.: Mikrochim. Acta 160 (1959)
353. HEGEDÜS A. J., NEUGEBAUER J., DVORSZKY M.: Mikrochim. Acta 282 (1959)
354. HEGEDÜS A. J., FUKKER K.: Mikrochim. Acta 357 (1962)
355. HEGEMANN F., PFAB B.: Glastech. Ber. **28,** 437 (1955)
356. HEGEMANN F., KOSTYROVA H., PFAB B.: Glastech. Ber. **30,** 14 (1957)
357. HEGEMANN F., HERT W.: Ber. deutsch. keram. Ges. **35,** 258 (1958)
358. HEGEMENN F., HERT W., SCHMIDT W.: Glastech. Ber. **31,** 81 (1958)
359. HEGEMANN F., SÜSS L.: Glastech. Ber. **31,** 185 (1958)
360. HEGEMANN F., Schmidt W., HERT W.: Glastech. Ber. **32,** 15 (1959)
361. HEGEMANN F., KÖSTER H. M., NEUBAUER G.: Ber. deutsch. keram. Ges. **37,** 483 (1960)
362. HEGEMENN F., OSTERRIED O.: Glastech. Ber. **33,** 201 (1960)
363. HEGEMENN F., OSTERRIED O.: Glastech. Ber. **33,** 285 (1960)
364. HEIDEL R., FASSEL V. A.: Anal. Chem. **23,** 784 (1951)
365. HEMINGWAY R. G.: Analyst **81,** 164 (1956)
366. HENDERSON E. H., OWERS M. J., WEBB M. S. W.: A.E.R.E. Report C/R – 2137, 1957
367. HERRMANN R., RICK W.: Naturwissen. **46,** 492 (1959)
368. HERRMANN R., LANG W.: Coll. Spectr. Inter. X., Lyon 1961
369. HERRMANN R., LANG W.: Z. ges. experim. Med. **134,** 268 (1961)
370. HERRMANN R., LANG W.: Optik **19,** 208 (1962)
371. HERRMANN R., LANG W.: Z. ges. exper. Med. **135,** 569 (1962)
372. HERRMANN R., LANG W.: Z. klin. Chem. **1,** 182 (1963)
373. HERRMANN R., ALKEMADE C. T. J., GILBERT P. T. (translator): Chemical Analysis by Flame Photometry, Interscience Publ. Wiley & Sons, New York, London 1963
374. HERTZ H.: Ann. Physik **31,** 421 (1887)
375. Hilger & Watts, Catalogue
376. HILL J. A.: Anal. Chem. **34,** 222 (1962)
377. HINNOV E.: J. Opt. Soc. Am. **47,** 151 (1957)
378. HINNOV E., KOHN H.: J. Opt. Soc. Am. **47,** 156 (1957)
379. HOFFMANN F. W., KOHN H.: J. Opt. Soc. Am. **51,** 512 (1961)

380. Holeyšová-Kozáková H.: Pracovní lékařství **11,** 311 (1959)
381. Honegger N.: Aetzl. Laboratorium **2,** 41 (1963)
382. Horn G. C.: Dissert. Univ. of Florida, Gainesville, U.S.A. 1956
383. Horr C. A.: Geological Survey Water Supply, Paper 14966
384. Horstman E. L.: Anal. Chem. **28,** 1417 (1956)
385. Hourigan H. F., Robinson J. W.: Anal. Chim. Acta **16,** 161 (1957)
386. Howling H. L., Landolt P. E.: Anal. Chem. **31,** 1818 (1959)
387. Hraško J. et. col.: Soil Analysis (in Slovakian) Slovakian Agricultural Publishing House, Bratislava 1962
388. Hübener H. J., Maurer H., Walter T.: Klin. Wochschr. **31,** 1095 (1953)
389. Huldt L., Lagerquist A.: Arkiv Fysik **2,** 333 (1950)
390. Hultgren R.: J. Am. Chem. Soc. **54,** 2320 (1932)
391. Humoller F. L., Walsh J. R., Wharton M. F.: J. Lab. Clin. Med. **48,** 127 (1956)
392. Humphrey J. R.: Anal. Chem. **37,** 1604 (1965)
393. Humpries E. C., v. Peach K., Trecey M. V.: Moderne Methoden der Pflanzenanalyse, p. 474, Part I., Springer, Berlin 1956
394. Ikeda S.: J. Chem. Soc. Japan, Pure Chem. Sect. **76,** 1122 (1955)
395. Ikeda S.: Sci. Rep. Res. Inst. Tokohu, Univ., Ser. A, **7,** 9 (1956)
396. Ikegami T., Matsuo T.: J. Japan Inst. Metals B, **14,** 51 (1950)
397. Ingamells C. O.: Talanta **9,** 681 (1962)
398. Inman W. R., Rogers R. A., Fournier J. A.: Anal. Chem. **23,** 483 (1951)
399. Intonti R. et al.: Rend. inst. super. sanite **20,** 100 (1957)
400. Ishida R.: J. Chem. Soc. Japan, Pure Chem. Sect. **76,** 56, 60 (1955)
401. Ishidate M., Mashiko Y., Kanroji Y.: J. Pharm. Soc. Japan **75,** 1492 (1955)
402. Ivanov D. N.: Pochvovedenie **7,** 423 (1949)
403. Ivanov D. N.: Pochvovedenie **11,** 61 (1953)
404. Ivanov D. N.: Zh. analit. Khim. **9,** 344 (1954)
405. Ivanov D. N.: Gidrokhim. Mater. **24,** 35 (1955)
406. Ivanov D. N., Kaplan B.: Zavodsk. Lab. **22,** 569 (1956)
407. Jackson P. J., Smith A. C.: J. Appl. Chem. **6,** 547 (1956)
408. Jackson W. P. U., Irwin L.: J. Clin. Pathol. **10,** 383 (1957)
409. James C. G., Sugden T. M.: Proc. Roy. Soc. A **227,** 312 (1955)
410. James C. G., Sugden T. M.: Nature **175,** 333 (1955)
411. Jansen H. W., Heyes J., Richter C.: Z. physik. Chem. A **174,** 291 (1935)
412. Jarman L., Manolitsis E., Matic M.: J. S. Afr. Inst. Min. Metallurgy **62,** 773 (1962)
413. Jentsch D., Jacob G.: Chem. Technik **7,** 93 (1955)
414. Jimeno Martin L., Sanchez Serano E.: Anales real Soc. espan. fíz. y quím. (Madrid), Ser. B, **47,** 175 (1951)
415. Johnson D. A., Lott P. F.: Anal. Chem. **35,** 1705 (1963)
416. Johnston B. R., Duncaon C. W., Lawton K., Benne E. J.: J. Assoc. Offic. Agr. Chemists **35,** 813 (1952)
417. Jones A. H.: Anal. Chem. **37,** 1761 (1965)

418. JONES D. I. H., TOMAS T. A.: Hilger J. **9,** 39 (1965)
419. JONES G. W., LEWIS B., FRIAUF J. B., PERROTT G. ST.: J. Am. Chem. Soc. **53,** 869 (1931)
420. JONES W. G., WALSH A.: Spectrochim. Acta **16,** 249 (1960)
421. JORDAN J. H. jun.: Petroleum Refiner **32,** 139 (1953)
422. JORDAN J. H. jun.: Petroleum Refiner **33,** 158 (1954)
423. JOST W.: Explosions und Verbrännungvorgänge in Gasen, Springer Verlag, Berlin 1939
424. JURČÍK F., ŠEBELA F.: Průmysl potravin **8,** 153 (1957)
425. JURY R. V., WEBB M. S. W., WEB R. J.: Anal. Chim. Acta **22,** 145 (1960)
426. KAISER H., SPECKER H.: Z. anal. Chem. **149,** 46 (1965)
427. KARCHMER J. H., GUNN E. L.: Anal. Chem. **24,** 1733 (1952)
428. KASHIMA J., MATUGACHI M.: Japan Analyst **4,** 420 (1955)
429. KASSNER J. L., BENSON V. M., CREITZ E. E.: Anal. Chem. **32,** 1151 (1960)
430. KEMULA W., BRACHACZEK W., DANCEWICZ D., HULANICKI A.: Chem. anal. (Warsaw) **3,** 729 (1958)
431. KHITRIN L. N.: Fysika goreniya i vzryva, Izdat. Moskov. Univ., Moscow 1957
432. KICK H.: Z. Pflanzenernähr., Düng. u. Bodenk. **67,** 53 (1954)
433. KICK H.: Z. anal. Chem. **151,** 406 (1956)
434. KICK H., BUCHER R.: Landwirtsch. Forsch. **10,** 96 (1957)
435. KICK H.: Z. Pflanzenernähr., Düng. u. Bodenk. **78,** 185 (1957)
436. KICK H.: Z anal. Chem. **163,** 252 (1958)
437. KICK H.: Z. Pflanzenernähr., Düng. u. Bodenk. **94,** 140 (1961)
438. KING A. S.: Astrophys. J. **27,** 353 (1908)
439. KINGSLEY W. K., WOLF G. E., WOLFRAM W. E.: Anal. Chem. **29,** 939 (1957)
440. KINSON K., HODGES R., BELCHER C. B.: Anal. Chim. Acta **29,** 134 (1963)
441. KINSON K., BELCHER C. B.: Anal. Chim. Acta **30,** 64 (1964)
442. KINSON K., BELCHER C. B.: Anal. Chim. Acta **31,** 180 (1964)
443. KIRILLOV A. I., ALKHIMENKOVA G. I.: Zavodsk. Lab. **31,** 57 (1965)
444. KLEMPERER R. L. VON: Thesis, Dresden Polytech. 1910
445. KNIGHT S. B., MATHIS W. C., GRAHAM J. R.: Anal. Chem. **23,** 1704 (1951)
446. KNIGHT S. B., PETERSON M. H.: Anal. Chem. **24,** 1514 (1952)
447. KNISELEY R. N., D'SILVA A. P., FASSEL V. A.: Anal. Chem. **35,** 910 (1963)
448. KNORRE G. F.: Combustion Processes (in Czech) SNTL Praha 1956
449. KNUTSON K. E.: Analyst **82,** 241 (1957)
450. KÖHLEIN J., LÜCKE K. E.: Z. Pflanzenernähr., Düng. u. Bodenk. **57,** 114 (1952)
451. KONOPICKY K., KAMPA P.: Tonind. Ztg. u. Keram. Rundschau **79,** 61 (1955)
452. KONOPICKY K., SCHMIDT W.: Z. anal. Chem. **174,** 262 (1960)
453. KONRAD H.: Nahrung **5,** 175 (1961)
454. KORTÜM G.: Kolorimetrie, Photometrie und Spektrometrie, Springer Verlag, Berlin 1955
455. KRAMER H.: Anal. Chim. Acta **17,** 521 (1957)
456. KRAMER H., PINTO L. J.: Anal. Chim. Acta **33,** 438 (1965)

457. KROPIK K.: Z. Pflanzenernahr,. Dung. u. Bodenk. **70,** 138 (1955)
458. KSANDOPULO G. J., SHCHERBOV D. P.: Zavodsk. Lab. **24,** 1432 (1958)
459. KUBIŠTA Z., KABICKÝ V., TIEZ N.: Hutn. listy **15,** 469 (1960)
460. KUBIŠTA Z., KABICKÝ V., TIEZ N.: Hutn. listy **17,** 813 (1962)
461. KUEMMEL D. F., KARL H. L.: Anal. Chem. **26,** 386 (1954)
462. KULINNA H.: Diplomarbet, Weimar, Silikat Technik **10,** 515 (1959)
463. KURITSKII A. L., MISIROVA E. P.: Nauch Tr. Tsentr. Nauch. Issled. Inst. Tsellyulozn. i Bumazhn. Prom. 161 (1961)
464. KUTSENKO Y. I.: Zh. analit. Khim. **13,** 107 (1958)
465. LAGERQUIST A., HULDT L.: Naturwissen. **42,** 365 (1955)
466. LANGE B.: Physik. Z. **31,** 139, 964 (1930)
467. LEBEDEV V. I.: Zh. analit. Khim. **14,** 283 (1959)
468. LEBEDEV V. I., VAINSTEIN E. E.: Zh. analit. Khim. **16,** 124 (1961)
469. LEBEDEV V. I.: Zh. analit. Khim. **16,** 272 (1961)
470. LEHMAN W.: Bodenkunde u. Pflanzenernahr. **9—10,** 766 (1938)
471. LEITHE W., HOFER A.: Mikrochim. Acta 268 (1961)
472. LEITHE W., HOFER A.: Microchim. Acta 277 (1961)
473. LENGYEL B., DOBOS S., TILL F.: Glastech. Ber. **33,** 206 (1960)
474. LEWIS B., PEASE R. N., TAYLOR H. S.: Combustion Processes, Princeton Univ. Press, Princeton 1956
475. LEYTON L.: Analyst **79,** 497 (1954)
476. LIEBIG J., BREDEHORST H.: Anal. Chim. Acta **24,** 573 (1962)
477. LOKEN H. F., TEAL J. S., EISENBERG E.: Anal. Chem. **35,** 875 (1963)
478. LOOMIS A. G., PERROTT G.: Ind. Eng. Chem. **20,** 1004 (1928)
479. LUH B. S., NIKETIC G.: Food res. **24,** 305 (1959)
480. LUNDEGARDH H.: Die quantitative Spektralanalyse von Elemente I., II., G. Fischer, Jena 1929, 1934
481. LUNDEGARDH H.: Metallwirtschaft **17,** 1222 (1938)
482. LUNDEGARDH H.: Die Blattanalyse, G. Fischer, Jena 1945
483. LUNDEGREN P.: J. Pharm. and Pharmacol. **5,** 511 (1953)
484. LYUTYI A. L., ROSSIKHIN V. S.: Inzh. Fiz. Zh. **3,** 101 (1960)
485. MACINTYRE I.: Rec. Trav. Chim. **74,** 498 (1955)
486. MACINTYRE I.: Biochem. J., **67,** 164 (1957)
487. MACKAY D. C., DELONG W. A.: Can. J. Agr. Sci. **34,** 451 (1954)
488. MAECK W. J., KUSSY M. E., GINTHER B. E., WHEELER G. V., REIN J. E.: Anal. Chem. **35,** 62 (1963)
489. MAGEE R. J., RAHMAN A. K. M.: Talanta **12,** 409 (1965)
490. MAHANTI P. S.: Phys. Rev. **42,** 609 (1932)
491. MALINOWSKI J., ROZMARYNOVICZ N.: Chem. anal. (Warsaw) **3,** 67 (1958)
492. MALINOWSKI J.: Chem. anal. (Warsaw) **4,** 939 (1959)
493. MALINOWSKI J., DANCEWICZ D.: Chem. anal. (Warsaw) **6,** 177 (1961)
494. MANDELSTAM S. L.: Compt. rend. acad. sci. U.S.S.R. **22,** 407 (1939)
495. MANNA L., STRUNK D. H., ADAMS S. L.: Anal. Chem. **29,** 1885 (1957)

496. MANNING D. C.: Atomic Absorp. Newsletter No. 25, p. 6, 1964
497. MANNING D. C.: Atomic Absorp. Newsletter **5**, 127 (1966)
498. MARCINKA K.: Chem. zvesti **13,** 479 (1959)
499. MARGOLIS L. D., KRAVCHENKO A. P., MUZICHENKO V. V.: Zavodsk. Lab. **28,** 1072 (1962)
500. MARGOSHES M., VALLEE B. L.: Anal. Chem. **28,** 180 (1956)
501. MARGOSHES M., VALLEE B. L.: Flame Photometry and Spectrometry, Principles and Applications, p. 353, in D. Glick: Methods of Biochemical Analysis, Vol. III., Interscience Publ., New York 1956
502. MARSH G. E.: Appl. Spectroscopy **12,** 113 (1958)
503. MARSHALL W. R. jun.: Chem. Eng. Progr. Monograph. p. 50, No. 2, (1954)
504. MARTENS P. H.: Chim. et ind. (Paris) **7,** 135 (1956)
505. MARTENS P. H.: Chim. anal. **39,** 361 (1957)
506. MASHIKO Y., KANROJI Y.: J. Pharm. Soc. Japan **76,** 441 (1956)
507. MASHIKO Y.: J. Pharm. Soc. Japan **76,** 1272 (1956)
508. MASON J. L.: Anal. Chem. **35,** 874 (1963)
509. MATHERS J. E., POTTER G. V., SHEARER N. W.: Anal. Chem. **30,** 1412 (1958)
510. MAVRODINEANU R., BOITEUX H.: L'analyse spectral quantitative par la flamme, Masson et Cie., Paris 1954
511. MAVRODINEANU R.: Spectrochim. Acta **17,** 1016 (1961)
512. MAZZAMARO P., TATOIAN G.: Anal. Chem. **26,** 1512 (1954)
513. MCBRIDE C. H.: U.S. At. Energy Comm. TID-7568, Pt 1, 213 (1958)
514. MCBRIDE C. H.: J. Assoc. Offic. Agr. Chemists **48,** 406 (1965)
515. MCCOY W. J., CHRISTIANSEN G. G.: Symposium on Flame Photometry, Spec. Tech. Publ. ASTM 116, 44 (1952)
516. MCCRACKAN M. L. et al.: Jour. A.O.A.C. **50,** 5 (1967)
517. MCPHERSON G. L., PRICE J. W., SCAIFE P. H.: Nature **199,** 371 (1963)
518. MEHLICH A., MONROE R. J.: J. Assoc. Offic. Agr. Chemists **35,** 588 (1952)
519. MEHLICH A., HARWARD M. E.: J. Assoc. Offic. Agr. Chemists **36,** 227 (1953)
520. MEHLICH A.: J. Assoc. Offic. Agr. Chemists **39,** 330 (1956)
521. MELOCHE V. W., BECK B. L.: Anal. Chem. **28,** 1890 (1956)
522. MENGOLI M., BOARI A.: Boll. Lab. Chim. Provinciali **9,** 341 (1958)
523. MENIS O.: U. S. At. Energy Co., Rep. CF – 52 – 8184 (1959)
524. MENIS O., RAINS T. C., DEAN J. A.: Anal. chem. **31,** 187 (1959)
525. METGE G.: Laboratoriumsbuch für Agrikulturchemiker, W. Knapp, Halle (Saale) 1951
526. MEVEL N., ANGOT J., VANOVERBERGHE L.: Chim. Anal. **42,** 15 (1960)
527. MITCHEL R. L.: Spectrochim. Acta **4,** 62 (1950)
528. MILOŠ J.: Chem. anal. (Warsaw) **7,** 373 (1962)
529. MIURA T.: Japan Analyst **14,** 310 (1965)
530. MIYAMOTO S. et al.: J. Japan Biochem. Soc. **27,** 632, 758 (1956)
531. MOBERG M. L., WAITHMAN V. B., ELLIS W. H., DUBOIS H. D.: Anal. Chem. **23,** 2053 (1951)

532. MORGAN M. E.: Atomic Absorp. Newsletter No. 21, p. 1 (1964)
533. MOSHER R. E., BIRD E. J., BOYLE A. J.: Anal. Chem. **22,** 715 (1950)
534. MOSHER R. E., ITANO M., BOYLE A. J., MYERS G. B., ISERI L. T.: Am. J. Clin. Pathol. **21,** 75 (1951)
535. MOSTYN R. A., CUNNINGHAM A. F.: Anal. Chem. **39,** 433 (1967)
536. MULLIN H. R., SHEEHY T. P.: Anal. Chem. **25,** 529 (1953)
537. MUSHA S., MUNEMORI M., NAKANISHI F.: Japan Analyst **13,** 330 (1964)
538. NAKAI T., ISHIDA R., HIDAKA S.: J. Chem. Soc. Japan, Pure Chem. Sect. **73,** 19 (1952)
539. NAZARENKO V. A., SHUSTOVA M. B., RAVITSKAYA V., NIKONOVA M. P.: Zavodsk. Lab. **28,** 537 (1962)
540. NEEB K. H., GEBAUHR W.: Z. anal. Chem. **162,** 167 (1958)
541. NEEB K. H.: Z. anal. Chem. **174,** 328 (1960)
542. NEEB K. H.: Z. anal. Chem. **184,** 414 (1961)
543. NEEB K. H.: Z. anal. Chem. **211,** 343 (1965)
544. NEUBERGER A., SCHÖFMANN E., HERKENHOFF K.: Arch. Eisenhuttenw. **29,** 35 (1958)
545. NICHOLSON J. L.: Appl. Spectroscopy **17,** 125 (1963)
546. NIKOLAEV G. I.: Zh. analit. Khim. **20,** 445 (1965)
547. NOVOBILSKÝ V., DVOŘÁK J.: Chem. listy **55,** 1049 (1961)
548. NUKIYAMA S., TANASAWA Y.: Trans. Soc. Mech. Engrs. **4,** 86, 138 (1938), **5,** 63 (1939)
549. OBOLENSKAYA L. I.: Pochvovedenie 420 (1951)
550. ODLER I., GEBAUER J.: Chem. zvesti **15,** 568 (1961)
551. OELSCHLAGER W.: Z. anal. Chem. **149,** 190 (1956)
552. OHASKI K. et al.: J. Chem. Soc. Japan, Ind. Chem. Sect. **57,** 16 (1954)
553. OLIVER J., GALBIS J. A.: Boll. inst. invest. agron. Madrid **16,** 255 (1956)
554. OLIVARI L., BENASSI R.: Boll. lab. chim. Provinciali, No. 4 (1959)
555. OLIVARI L., BENASSI R.: Boll. lab. chim. Provinciali **11,** 330 (1960)
556. OSBORN G. H., JOHNS H.: Analyst **76,** 410 (1951)
557. OUNSTED D.: J. Inst. Fuel **31,** 474 (1958)
558. OVCHAR L. A., VITKUN R. A., POLUEKTOV N. S.: Zh. analit. Khim. **20,** 554 (1965)
559. PAPPAS E. G., ROSENBERG L. A.: Journ. A.O.A.C. **49,** 792 (1966)
560. PARKER H. E.: Atomic Absorp. Newsletter No 13, p. 1 (1963)
561. PASSMORE W. O., ADAMS P. B.: Atomic Absorp. Newsletter **5,** 77 (1966)
562. PATROVSKÝ V.: Coll. Czech. Chem. Commun. **26,** 2445 (1961)
563. PAWLOWSKA H., ZLOTOWSKA Z.: Mater. budow. **9,** 196 (1954)
564. PEARSE R. W. B., GAYDON A. G.: The Identification of Molecular Spectra, Chapman and Hall, Ltd., London 1950
565. PERKIN—ELMER: Catalogue
566. PERKIN—ELMER: Atomic absorption spectrophotometer — Catalogue
567. PERKINS J.: Analyst **88,** 324 (1963)

568. PETERSON G. E.: Atomic Absorp. Newsletter **5,** 142 (1966)
569. PHIFER L. H.: Anal. Chem. **29,** 1528 (1957)
570. PIETZKA G., CHUN H.: Angew. Chem. **71,** 276 (1959)
571. PILGRIM W. E., FORD W. R.: Anal. Chem. **35,** 1735 (1963)
572. PINTA M.: J. recherches centre natl. recherche sci. labs. Bellevue, Paris, No. 21, 260 (1952)
573. PINTA M.: Thesis, Univ. Paris, 1952
574. PINTA M.: Chim. anal. **36,** 126 (1954)
575. PINTA M., AUBERT H.: Compt. rend. **244,** 873 (1957)
576. PIPER C. S.: Soil and Plant Analysis, Interscience Publ., Inc., New York 1944
577. PIPER E., HAGEDORN H.: Arch. Eisenhüttenw. **22,** 299 (1951)
578. PLATTE J. A., MARCY V. M.: Atomic Absorp. Newsletter **4,** 289 (1965)
579. PLŠKO E.: Chem. zvesti **16,** 762 (1962)
580. POLUEKTOV N. S., KONONENKO L. I., NIKONOVA M. P.: Zh. analit. Khim. **12,** 10 (1957)
581. POLUEKTOV N. S., NIKONOVA M. P., LEIDERMAN T. A., LAUER G. S.: Zh. analit. Khim. **12,** 699 (1957)
582. POLUEKTOV N. S.: Ekspresnyje metody analiza pri pomošči fotometrii plameni v cvetnoj metallurgii (Rapid flame-photometric analytical methods in non-ferrous metallurgy), Metallurgizdat, Moscow 1958
583. POLUEKTOV N. S., NIKONOVA M. P.: Zavodsk. Lab. **24,** 528 (1958)
584. POLUEKTOV N. S., NIKONOVA M. P., VITKUN A. A.: Zh. analit. Khim. **13,** 48 (1958)
585. POLUEKTOV N. S.: Metody analiza po fotometrii plameni (Flame-photometric analytical methods) Goschimizdat, Moscow 1959
586. POLUEKTOV N. S., NIKONOVA M. P.: Ukr. Khim. Zh. **25,** 217 (1959)
587. POLUEKTOV N. S., KONONENKO L. I., NIKONOVA M. P.: Zh. analit. Khim. **12,** 10. (1959)
588. POLUEKTOV N. S., NIKONOVA M. P., GRINZAJD S. E.: Zavodsk. Lab. **26,** 161 (1960)
589. POLUEKTOV N. S., POPOVA S. B., OVCHAR L. A.: Zh. analit. Khim. **15,** 131 (1960)
590. POLUEKTOV N. S., POPOVA S. B.: Zh. analit. Khim. **15,** 437 (1960)
591. POLUEKTOV N. S., VITKUN R. A.: Zh. analit. Khim. **16,** 260 (1961)
592. POLUEKTOV N. S.: Zavodsk. Lab. **28,** 1069 (1962)
593. POLUEKTOV N. S., VITKUN R. A.: Zh. analit. Khim. **17,** 935 (1962)
594. POLUEKTOV N. S., MESHKOVA S. B., NIKONOVA M. P.: Zavodks. Lab. **30,** 563 (1964)
595. POMPOWSKI T., WIEWIOROWSKI E.: Chem. anal. **4** (Warsaw) 487 (1959)
596. POPOV D. K., MIKHAILOVA A. I.: Zh. analit. Khim. **17,** 440 (1963)
597. PORTER P., WYLD G.: Anal. Chem. **27,** 733 (1955)
598. PORTER P., WYLD G.: Anal. Chem. **27,** 733 (1955)
599. POSSIDONI DE ALBINATI J. F., CAPACCIOLI J. H.: Annales Assoc. quim. Arg.: **42,** 115 (1954)
600. POSSIDONI DE ALBINATI J. F.: Annales Assoc. quim. Arg. **43,** 106 (1955)

601. Poulos P. P., Pitts R. F.: J. Lab. Clin. Med. **49,** 300 (1957)
602. Powell F. J. N.: J. Clin. Pathol. **6,** 286 (1953)
603. Powell R. W.: Trans. Farad. Soc. **39,** 311 (1943)
604. Powell R. J., Todd J.: Trans. Soc. Glass. Technol. **43,** 73 T (1959)
605. Pro M. J., Nelson R. A., Mathers A. P.: J. Assoc. Offic. Agr. Chemists **39,** 506 (1956)
606. Pro M. J., Nelson R. A.: J. Assoc. Offic. Agr. Chemists **40,** 575 (1957)
607. Proks J., Plško E., Obert T.: Chem. zvesti **17,** 829 (1963)
608. Prudnikov E. D.: Zhur. anal. chim. **20,** 40 (1965)
609. Prudnikov E. D.: Zhur. anal. chim. **20,** 1248 (1965)
610. Prytz B.: Bull. St. Francis Sanatorium (Roslyn) **10,** (3), 27 (1953)
611. Puffeles M., Nessim N. E.: Analyst **82,** 467 (1957)
612. Puffeles M., Nessim N. E.: Anal. Chim. Acta **20,** 38 (1959)
613. Pungor E., Hegedüs A. J.: Mag. Kém. Foly. **61,** 308 (1955)
614. Pungor E., Zapp E. E.: Mag. Kém. Foly. **63,** 188 (1957)
615. Pungor E., Konkoly-Thegé F.: Microchim. Acta 712 (1959)
616. Pungor E., Zapp E. E.: Annales Univ. Sci. Budapest. de R. Eötvös nom. sect. chim. tom II. 451 (1960)
617. Pungor E., Zapp E. E.: Magyar Kém. Foly. **66,** 523 (1960)
618. Pungor E., Wezprémy B., Pályi M.: Microchim. Acta 436 (1961)
619. Pungor E., Mahr M.: Talanta **10,** 537 (1963)
620. Püschel A., Eckhard S.: Arch. Eisenhüttenw. **30,** 731 (1959)
621. Radmacher W., Schmitz W.: Brennstoff-Chem. **38,** 270 (1957)
622. Rains T. C., House H. P., Menis O.: Anal. Chim. Acta **22,** 315 (1960)
623. Rains T. C., Zittel H. E., Ferguson M.: Anal. Chem. **34,** 778 (1962)
624. Rains T. C., Zittel H. E., Ferguson M.: Talanta **10,** 367 (1963)
625. Ramakrishna T. V., Robinson J. W., West P. W.: Anal. Chim. Acta **37,** 20 (1967)
626. Ranz W. E., Marshall J. R.: Chem. Eng. Progr. **48,** 142 (1952)
627. Rathje A. O.: Paper presented at the Meeting of the Am. Chem. Soc. (Sept. 7th—10th) (1953)
628. Ratner R., Scheiner D.: Analyst **89,** 136 (1964)
629. Rawling B. S., Greaves M. C., Amos M. P.: Nature **188,** 137 (1960)
630. Reed M. G., Scott A. D.: Anal. Chem. **33,** 773 (1961)
631. Reich H., Grabbe F.: Tonind.-Ztg. **79,** 124 (1955)
632. Reissig W.: Kleines agrikultur-chemische Praktikum, Deutscher Bauernverlag, Berlin 1956
633. Řezáč Z., Dvořák J.: Z. anal. Chem. **174,** 96 (1960)
634. Řezáč Z., Dvořák J.: Acta Chim. Acad. Sci. Hung. **30,** 375 (1962)
635. Řezáč Z.: Lecture presented at a meeting of the Spectral Analysis Research Association, 1963
636. Řezáč Z., Dvořák J.: Chem. prům. **14,** 485 (1964)
637. Řezáč Z.: not published

638. RIEHM H.: Bodenkunde u. Pflanzernähr. **36,** 109 (1945)
639. RIETHMÜLLER H. U.: Mikrochim. Acta 178 (1953)
640. ROBINSON A. M., OVENSTON T. C. J.: Analyst **79,** 47 (1954)
641. ROBINSON G. A.: Canad. J. Biochem. Physiol. **38,** 643 (1960)
642. ROBINSON J. W.: Anal. Chim. Acta **23,** 479 (1960)
643. ROBINSON J. W.: Anal. Chim. Acta **24,** 451 (1960)
644. ROBINSON J. W.: Anal. Chim. Acta **24,** 254 (1961)
645. ROBINSON J. W., KEVAN L. J.: Anal. Chim. Acta **28,** 170 (1963)
646. ROBINSON J. W., DEKKER M.: Atomic Absorption Spectroscopy, New York 1966
647. RODDEN C. J., PLANTINGA O. S.: Phys. Revs. **45,** 280 (1934)
648. ROSEN B., WENIGER S.: Advances in Molecular Spectroscopy, Pergamon Press 1962
649. ROTHE C. F., SAPIRSTEIN L. A.: Am. J. Clin. Pathol. **25,** 1076 (1955)
650. ROTHE C. F., SAPIRSTEIN L. A.: Techn. Bull. Registry Med. Technologists **25,** 184 (1955)
651. ROTHERMEL O. L., NORDBERG M. B.: Am. Ceram. Soc. Bull. **31,** 324 (1952)
652. ROUSSELET F.: Afinidad **22,** 185 (1965)
653. ROY N.: Anal. Chem. **28,** 34 (1956)
654. RUBEŠKA I.: Z. anal. Chem. **186,** 300 (1962)
655. RUBEŠKA I., MOLDAN B.: Záv. zpráva UÚG, Praha 7/7 (1962)
656. RUBEŠKA I., MOLDAN B., VALNÝ Z.: Anal. Chim. Acta **29,** 206 (1963)
657. RUBEŠKA I., MOLDAN B.: Rudy **12,** 191 (1964)
658. RUBEŠKA I., MOLDAN B.: Chem. listy **59,** 1119 (1965)
659. RUBEŠKA I., MOLDAN B.: Coll. Czech. Chem. Commun. **30,** 1731 (1965)
660. RUBEŠKA I., MOLDAN B.: Acta chim. Acad. Sci. Hung. **44,** 367 (1965)
661. RUBEŠKA I., ŠULCEK Z., MOLDAN B.: Anal. chim. Acta **37,** 27 (1967)
662. RUTKOWSKI W., MALINOWSKI J.: Chem. anal. (Warsaw) **6,** 1065 (1961)
663. SAHA M. N.: Phil. Mag. **40,** 472 (1920)
664. SACHDEV S. L., ROBINSON J. W., WEST P. W.: Anal. Chim. Acta **38,** 499 (1967)
665. SAUMAN Z.: Chem. zvesti **11,** 168 (1957)
666. SAUTER L.: Forsch. Geb. Ingenieurwesens No. 279, 1926
667. SCHACHTSCHABEL P., SCHWERTMANN U.: Z. Pflanzenernähr. Düng. Bodenk. **82,** 38 (1958)
668. SCHALL E. D., HAGELBERG R. R.: J. Assoc. Offic. Agr. Chemists **35,** 757 (1952)
669. SCHAPIRO L., BRANNOCK W.: Sprechsaal Keram., Glass, Email, **88,** 188 (1955)
670. SCHARRER K., MENGEL K.: Landwirtsch. Forsch. **9,** 204 (1956)
671. SCHARRER K., MENGEL K.: Z. Pflanzenernähr. Düng. Bodenk. **77,** 18 (1957)
672. SCHARRER K., MENGEL K.: Z. Tierphysiol. Tierernähr. u. Futtermittelk. **13,** 142 (1958)
673. SCHINKMANN A.: Silikat-Tech. **2,** 163 (1951)
674. SCHLÄTER R.: Spectrochim. Acta **11,** 361 (1956)
675. SCHLÜTZ G. O.: Z. physiol. Chem. **293,** 254 (1953)
676. SCHMAUCH G. F., SERFASS E. J.: Anal. Chem. **30,** 1160 (1958)

677. Schmid A., Zipf K.: Biochem. Z. **331,** 144 (1959)
678. Schmid A., Zipf K.: Biochem. Z. **333,** 84 (1960)
679. Schmitz W.: Jahresber. Limol. Flusstat Freudenthal **45** (1950)
680. Schneider R.: Landwirtsch. Forsch. **6,** 200 (1954)
681. Schreiber E.: Z. anal. Chem. **210,** 93 (1965)
682. Schrenk W. G., Graber K., Johnson R.: Anal. Chem. **33,** 106 (1961)
683. Schroeder D.: Z. Pflanzenernähr. Düng. Bodenk. **73,** 86 (1956)
684. Schuhknecht W.: Angew. Chem. **59,** 299 (1937)
685. Schuhknecht W. S., Schinkel H.: Brennstoff-Chem. **38,** 275 (1957)
686. Schuhknecht W., Schinkel H.: Z. anal. Chem. **160,** 23 (1958)
687. Schuhknecht W.: Die Flammenspektralanalyse, F. Enke, Stuttgart 1961
688. Schuhknecht W., Schinkel H.: Brennstoff-Chem. **42,** 292 (1961)
689. Schuhknecht W., Schinkel H.: Z. anal. Chem. **194,** 161 (1963)
690. Schüler V. C. O., Jansen A. V., James G. S.: J. Afr. Inst. Min. Metall., **62,** 807 (1962)
691. Scott R. K., Marcy V. M., Hronas J. J.: Symposium on Flame Photometry, ASTM Spec. Tech. Publ. **116,** 105 (1951)
692. Scott T. C., Roberts E. D., Cain D. A.: Atomic Absorp. Newsletter **6,** 1 (1967)
693. Serfass E. J.: Bull. dsE 1819 Pensalt Chemicals Corp., Philadelphia
694. Shafto R. G.: Atomic Absorp. Newsletter **3,** 115 (1964)
695. Shaw W. M., Veal N. C.: Proc. Soil Sci. Soc. Am. Proc. **20,** 328 (1956)
696. Shaw W. M.: Anal. Chem. **30,** 1682 (1958)
697. Shaw W. M.: J. Agr. Food Chem. **9,** 18 (1961)
698. Shellenberg T. E., Pyke R. E., Parrish D. B., Schrenk W. G.: Anal. Chem. **32,** 210 (1960)
699. Shionoya M., Asano C.: Japan Analyst **5,** 384 (1956)
700. Shoji K., Nakata S.: Proc. Am. Soc. Hort. Sci. **64,** 299 (1952)
701. Shiryaeva O. A., Melamed S. G.: Zavodsk. Lab. **30,** 183 (1964)
702. Skewes H. R.: Proc. Austral. Inst. Min. Metall. 217 (1964)
703. Slavin W., Manning D.: Anal. Chem. **35,** 253 (1963)
704. Slavin W., Sprague S.: Atomic Absorp. Newsletter No. 17, **1** (1964)
705. Sprague S., Rieders F., Cordova V.: Atomic Absorp. Newsletter No. 17, 7 (1964)
706. Slavin W.: Atomic Absorp. Newsletter No. 24, 15 (1964)
707. Slavin W., Venghiattis A., Manning D. C.: Atomic Absorp. Newsletter **5,** 84 (1966)
708. Smales A. A.: Analyst **76,** 348 (1951)
709. Smit J. A., Vendrik A. J. H.: Physica **14,** 505 (1948)
710. Smith F. M.: V. S. Atomic Energy Comm. Rep. HW-59864, 1959
711. Smith G. W., Palmby A. K.: Anal. Chem. **31,** 1798 (1959)
712. Smith H., Sugden T. M.: Proc. Roy. Soc. A 211, 58 (1952)
713. Smith W.: Amer. Sci. **5,** 301 (1873)

714. Snelleman W., Smit J. A.: Colloquium Spetroscopicum Internationale VI, Pergamon Press, London 1957, 44
715. Sokolovski A. A.: Zh. Khim. Prom. **13,** 92 (1936)
716. Sommer H.: Monatsh. Veterinärmed. **11,** 154 (1956)
717. Southern Analytical: A 1700 Atomic Absorption Spectrometer catalogue
717a. Specker H.: Z. Erzbergb. Metallhüttenw. **17,** 132 (1964)
718. Spencer R. P., Mitchel T. G., King E. R.: J. Lab. Clin. Med. **50,** 646 (1957)
719. Spindler F., Wolf E. F.: Landwirtsch. Forsch. **9,** 179 (1956)
720. Sprague S., Slavin W.: Atomic Absorp. Newsletter No. 12, 4 (1963)
721. Sprague S.: Atomic Absorp. Newsletter No. 13, 8 (1963)
722. Sprague S.: Atomic Absorp. Newsletter No. 14, 9 (1963)
723. Sprague S., Manning D. C., Slavin W.: Atomic Absorp. Newsletter No. 20, 1 (1964)
724. Sprague S., Slavin W.: Atomic Absorp. Newsletter No. 20, 11 (1964)
725. Spurný K., Jech Č., Sedláček B., Štorch O.: Aerosols (in Czech) SNTL Praha 1961
726. Stamicarbon N. V.: Brit. Pat. 779, 212, 19. 4. 1955
727. Standen G. W., Tennant C. B.: Anal. Chem. **28,** 858 (1956)
728. Stavinoha W. B., Nash J. B.: Anal. Chem. **32,** 1965 (1960)
729. Stewart J. M., McIlhenny R. C.: U.S. Atomic Energy Comm. Rep. Y – 1140 (1956)
730. Stewart W. K., Hutchinson F., Fleming L. W.: J. Lab. Clin. Med. **61,** 858 (1963)
731. Stone M., Thomas J. E.: Analyst **83,** 691 (1958)
732. Strasheim A., Nell J. P.: J. S. Afr. Chem. Inst. **7,** 79 (1954)
733. Strasheim A., Strelow F. W. E., Butler L. R. P.: J.S. Afr. Chem. Inst. **13,** 73 (1960)
734. Strasheim A., Butler L. R. P., Maskew E. C.: J.S. Afr. Inst. Min. Metall. **62,** 796 (1962)
735. Strasheim A., Norval E., Butler L. R. P.: J.S. Afr. Chem. Inst. **17,** 55 (1964)
736. Straubel H.: Naturwissen. **40,** 337 (1953)
737. Straubel H.: Mikrochim. Acta 329 (1955)
738. Stumpf K. E., Consier T.: Colloquium Spectroscopicum Internationale VI, Pergamon Press, London 1957, p. 35
739. Stumpf K. E., Consier T.: Colloquium Spectr. Inter. IX. Lyon 1961
740. Štupar J.: Nuklearni inst. J. Stefan, NIJS Report R-425, 1964
741. Sugawara K., Koyama T., Kawasaki N.: Bull. Chem. Soc. Japan **29,** 679 (1956)
742. Sugawara K., Koyama T., Kawasaki N.: Bull. Chem. Soc. Japan **29,** 683 (1956)
743. Sugden T. M., Knewstubb P. F.: Research Correspondence **9,** 32 (1956)
744. Suhr N. H., Ingamells C. O.: Anal. Chem. **38,** 730 (1966)
745. Suter H., Hadorn H.: Mitt. Lebensmitt. Hyg. (Bern) **51,** 107 (1960)

746. SYKES P. W.: Analyst **81,** 283 (1956)
747. TAPPE W., CALKER J. VAN: Z. Anal. Chem. **198,** 13 (1963)
748. TAKEUCHI T., SUZUKI M.: Talanta **11,** 1391 (1964)
749. TELOH H. A.: Clinical Flame Photometry, Thomas. Springfield Ill., 1959
750. THILLIEZ G.: Anal. Chem. **39,** 427 (1967)
751. THOMAS R. D.: Anal. Chem. **38,** 785 (1966)
752. THUN R., HERRMANN R., KNICKMANN E.: Handbuch der landwirtschaftlichen Versuchs- und Untersuchungsmethodik (Methodenbuch, Vol. I., Die Untersuchung von Boden, 3 Ed., Neumann, Berlin 1955)
753. TODD H. E., TRAMUTT H. M.: Anal. Chem. **26,** 1137 (1954)
754. TORIBARA T. Y., DEWEY P. A., WARNER H.: Anal. Chem. **29,** 540 (1957)
755. TÖRÖK T.: Z. anal. Chem. **119,** 120 (1940)
756. TOTH S. J., PRINCE A. L.: Soil. Sci. **67,** 439 (1949)
757. TOYAMA I., TAKAUCHI K.: Japan Analyst **9,** 123 (1960)
758. TRENT D., SLAVIN W.: Atomic Absorp. Newsletter **3,** 131 (1964)
759. TRENT D., SLAVIN W.: Atomic Absorp. Newsletter **4,** 300 (1965)
760. TRUNEČEK V.: Čs. časopis pro fyziku **10,** 191 (1960)
761. TSKHAI N. S.: Opt. i spektroskopiya **12,** 524 (1962)
762. TUCHOLKA-SZMEJA B.: Chem. anal. (Warsaw) **1,** 255 (1956)
763. TŮMA J.: Silikáty **5,** 333 (1961)
764. TYLES J. B.: Atomic Absorp. Newsletter **6,** 14 (1967)
765. U.K.A.E.A. Report PG 59 (S), 1962
766. UNGER J., UNGER L.: Glastech. Ber. **29,** 15 (1956)
767. Unicam SP900 – catalogue
768. Unicam SP90 – catalogue
769. UZUMASA Y.: J. Chem. Soc. Japan, Pure Chem. Sect. **79,** 1292 (1958)
770. UZUMASA Y. et al.: J. Chem. Soc. Japan, Pure Chem. Sect. **81,** 430 (1960)
771. VAINSTEIN E. E., LEBEDEV V. I.: Zh. analit. Khim. **16,** 670 (1961)
772. VALENCIA R.: Bull. Soc. Chim. Biol. **38,** 1071 (1956)
773. VALLEE B. L., BAKER M. R.: Pittsburgh Conference on Anal. Chem. and Appl. Spectroscopy, March 1955
774. VALLEE B. L., BARTHOLOMAY A. F.: Anal. Chem. **28,** 1753 (1956)
775. VALORI P., SAVOINI F.: Ricerca Sci. **27,** 791 (1957)
776. VALORI P., SAVOINI F.: Ricerca Sci. **27,** 1204 (1957)
777. VALORI P., TALENTI M., SAVOINI F.: Ricerca Sci. **27,** 1901 (1957)
778. VALORI P., ALASIA A. M., SAVOINI F.: Ricerca Sci. **28,** 1004 (1958)
779. VEITZ J. V., GOUVICH L. V.: Opt. i Spektroskopiya **1,** 22 (1956)
780. VEITZ J. V., GOURVICH L. V.: Opt. i Spektroskopiya **2,** 274 (1957)
781. VELTISHCHEV Y. E., ZLATKOVSKAYA N. M., FELDMAN M. G.: Lab. Delo **7,** 6 (1961)
782. VOINOVITCH I. A., DEBRAS J.: Bull. soc. franc. céram. No. 32, 29 (1956)
783. WADE M. A., SEIM H. J.: Anal. Chem. **33,** 793 (1961)
784. WALLACE F. J.: Hilger J. **7,** 39 (1962)

785. WALLACE F. J.: Analyst **88,** 259 (1963)
786. WALLACE F. J.: Hilger J. **7,** 65 (1963)
787. WALSH A.: Spectrochim. Acta **7,** 108 (1955)
788. WARREN R.: Proc. VIII. Colloquium Spectr. Internationale p. 213, Sauerländer, Aarau 1960
789. WARREN R. L.: Analyst **90,** 549 (1965)
790. WATANABE H., KENDALL K. K. jun.: Appl. Spectroscopy **9,** 132 (1955)
791. WEBB M. S. W., WORDINGHAM M. L.: Anal. Chim. Acta **28,** 450 (1963)
792. WEHNER G., BUNGE W.: Chem. Tech. **5,** 251 (1953)
793. WEISS J., BIEBER B.: Hutn. listy **14,** 247 (1959)
794. WERNER W.: Landwirtsch. Forsch. **13,** 273 (1960)
795. WEST A. C., COOK W. D.: Anal. chem. **32,** 1471 (1960)
796. WEST A. C.: Anal. chem. **36,** 310 (1964)
797. WEST P. W., FOLSE P., MONTGOMERY D.: Anal. chem. **22,** 667 (1950)
798. WESTERHOFF H.: Landwirtsch. Forsch. **7,** 128 (1955)
799. WEVER F., KOCH W., WIETHOFF D.: Arch. Eisenhuttenw. **24,** 383 (1955)
800. WILDY P. C.: A.E.R.E. Report CIR 2114, 1956
801. WILKANOWICZ M.: Bull. Inst. Techn. Krzemianów **3,** 1 (1954); **5,** 150 (1954)
802. WILL E. G., SCHWARZKOPF B.: J. Am. Water Works Assoc. **47,** 53 (1955)
803. WILLGALLIS A.: Z. anal. Chem. **157,** 249 (1957)
804. WILLIAMS C. H.: Anal. Chim. Acta **22,** 163 (1960)
805. WILLIAMS C. H., DAVID D. J., IISMAA O.: J. Agric. Sci. **59,** 381 (1962)
806. WILLIAMS J. P., ADAMS P. B.: J. Am. Ceram. Soc. **37,** 306 (1955)
807. WILLIAMS T. R., MORGAN R. R. T.: Chemy Ind. (No. 37), 970 (1953)
808. WILLIS J. B.: Nature **184,** 186 (1959)
809. WILLIS J. B.: Spectrochim. Acta **16,** 259 (1960)
810. WILLIS J. B.: Spectrochim. Acta **16,** 273 (1960)
811. WILLIS J. B.: Spectrochim. Acta **16,** 551 (1960)
812. WILLIS J. B.: Anal. Chem. **33,** 556 (1961)
813. WILLIS J. B.: Nature **192,** 929 (1961)
814. WILLIS J. B.: Anal. Chem. **34,** 614 (1962)
815. WILLIS J. B.: Nature **207,** 715 (1965)
816. WILSON H. W.: Anal. Chem. **38,** 920 (1966)
817. WILSON L.: Anal. Chim. Acta **30,** 377 (1964)
818. WILSON L.: Anal. Chim. Acta **35,** 123 (1966)
819. WILSON R. H., CONWAY M. B.: Natl. Bur. Standards Circ. **523,** 111 (1954)
820. WINEFORDNER J. D., LATZ H. W.: Anal. Chem. **33,** 1727 (1961)
821. WINEFORDNER J. D., VICKERS T. J.: Anal. Chem. **36,** 161 (1964)
822. WINEFORDNER J. D., STAAB R. A.: Anal. Chem. **36,** 165 (1964)
823. WINTER K. A.: Z. ges. inn. Med. Grenzgebiete **11,** 324 (1956)
824. WOLFE H. C.: Temperature, Its Measurement and Control in Science and Industry, Reinhold Publ. Co., New York 1955
825. WOODLAND D. J., MACH A.: J. Am. Chem. Soc. **55,** 3149 (1933)

826. WÜNSCH A., TEICHER K.: Z. Pflanzenernähr. Düng. Bodenk. **97,** 101 (1962)
827. YAMASAKI T., KUSANO S.: Bull. Div. Pl. Breed. Cult., Tokai–Kinki Natn. Agric. Exp. Stn. **3,** 1 (1956)
828. YESIKOV A. D., BEZSHCHASTNOVA G. S., YAKOVLEV G. M.: Izv. Acad. Sci. USSR, Geol. ser. **24,** 69 (1959)
829. YOE H. J., KOCH H. R. jun.: Trace analysis, Wiley & Sons, New York 1957
830. YOFÉ J., FINKELSTEIN R.: Anal. Chim. Acta **19,** 166 (1958)
831. YOFÉ J., STILLER M.: Bull. Res. Council Israel A, **11,** 132 (1962)
832. YOFÉ J., AVNI R., STILLER M.: Israel Atomic Energy Commission I A – 770 (1962)
833. YOFÉ J., AVNI R., STILLER M.: Anal. Chim. Acta **28,** 331 (1963)
834. YOUDEN W. J.: Anal. Chem. **32,** 23A (1960)
835. YOSHIZAKI T.: Anal. Chem. **35,** 2177 (1963)
836. ZAGRODZKI S., ZAORSKA H.: Die Lebensmittel-Industrie **5,** 194 (1958)
837. ZAIDMAN N., ORECHKIN D.: Novosti Neftyanoi Tech. USSR **5,** 25 (1955)
838. ZAKHARIYA N. F., FUGA N. A., LEIDERMAN C. A.: Zavodsk. Lab. **22,** 1303 (1956)
839. ZAKHARIYA N. F., FUGA N. A.: Fiz. Sb. l'vov. gosud. Univ. IV. 355 (1958)
840. ZAKHARIYA N. F., LEIDERMAN C. A.: Fiz. Sb. l'vov. gosud. Univ. IV. 358 (1958)
841. ZEEMAN P. B., BUTLER L. R. P.: Appl. Spectroscopy **16,** 120 (1962)
842. Zeiss Carl, Catalogue
843. ZELUKOVA Y. V., POLUEKTOV B. S.: Zh. analit. Khim. **18,** 435 (1963)
844. ZELUKOVA Y. V., NIKONOVA M. P., POLUEKTOV N. S.: Zh. analit. Khim. **21,** 1409 (1966)
845. ZETTNER A., SELIGSON D.: Clin. Chem. **10,** 869 (1964)
846. ZHITKEVICH V. F. et al.: Izv. Vyssh. ucheb. Zavedenii Fizika (Tomsk) **78** (1963)
847. ZHITKEVICH V. F., LYUTYI A. J., NESTERKO N. A., ROSSIKHIN V. S., TSIKORA J. L.: Optika Spektrosk. **14,** 35 (1963)
848. ZHITKEVICH V. F., LYUTYI A. I., NESTERKO N. A., ROSSIKHIN V. S., TSIKORA J. L.: Optika Spektrosk. **14,** 336 (1963)
849. ZHURAVLEV G. J., GAVRILOVA I. A.: Zh. analit. Khim. **19,** 54 (1964)

INDEX

Absorbance, 32, 33
Absorption coefficient, 30–32, 35
Absorption equation, 31
Absorption flame photometry, 14–15, 17, 23, 28, 202, 203
  classification of methods, 19–20
  interfering influences, 166
  operating technique, 156–58
Absorption spectra of elements, table of, 216
Accuracy, 158–62
Acetylene, 67–69, 111, 129
  see also Flames
Acetylene pressure, 130–31
Aerosols, 51–53, 77
  coagulation, 57
  droplet evaporation in mist chamber, 58–61
  droplet sedimentation, 57
  droplet size, 61, 77
    and size distribution, 53–57
  fineness of, 55
  mean particle size of, 79
Agricultural chemistry, 269–72
Agrochemical analyses, 269
Air, 110
  as oxidant, 65–67
Air pressure, 110, 130
Alkali halides, 181, 182
Alkali metal oxides, 253
Alkali metals, 13, 15, 21, 117, 139, 141, 181, 183, 192, 213, 266, 269
Alkaline earth elements, 13, 15, 21, 27, 49, 62, 173, 174, 175, 184, 191, 197, 219
Aluminium, 174, 176, 178, 180, 228, 229, 250
Aluminium nitrate, 176, 239, 241, 245, 246
Aluminium oxide, 176
Ammonium phosphate, 195, 260
Analysis of materials, literature review, 280
Analytical methods, literature citations, 280
Analytical results
  data recorded, 144
  determination of, 143–48
  preprinted forms for, 144
  reading from calibration curve, 145
Angular dispersion, 94
Anion addition, 191
Antimony, 232
Applications, 236–74
  agricultural chemistry, 269–72
  biochemistry, 272–73
  chemical industry, 245–59
  literature review, 280
  miscellaneous, 273–74
  water analysis, 237–45
Arsenic, 232
Atomic absorption, 14, 20, 23, 114
Atomic fluorescence, 15, 20
Atomic spectra, 26, 27
Automatic analysers, 274

Back-flashing, 112
Band, 29
Band spectrum, 16

Barium, 12, 184, 219, 223
BECKMAN burner, 74, 78, 115
  modification of, 76
BECKMAN spectrophotometer, 114
Biochemistry, 272–73
Bismuth, 233
Blank experiment, 164, 165
Blank experiment value, 164
Blocking, 175
Blood serum, 272
Boiling points, 174–75
Boltzmann constant, 33
Boltzmann equation, 200
Boltzmann's law, 33
Boric acid, 177
Boron, 21, 177, 228
Bracketing method, 148
Buffer solutions, 197–98, 243, 250, 252
BUNSEN burner, 11, 15, 41–42, 73
Burners, 40–43, 48, 49, 72–77, 112, 122
  adjusting, 129–30, 132, 134
  BECKMAN, 74, 78, 115
    modification of, 76
  BUNSEN, 11, 15, 41–42, 73
  butterfly-type, 73, 268
  direct-injection, 52, 57, 61, 72, 74–75, 78, 80, 82, 114, 122, 124, 169, 179, 184, 199, 201
  flat, 132
  maintenance, 137–38
  MECKER, 43, 73, 112, 122, 132
  miscellaneous types, 75
  positioning of, 201
  slit-type, 122, 125
  special types, 75
  water-cooled, 73
  with mist chambers in inlet, 72
Butane, 69–70

Cadmium, 117, 226, 227
Caesium, 21, 83, 117, 155, 189, 193, 215, 262
Calcium, 12, 23, 119, 146–47, 171, 173, 174, 176–78, 180, 184, 195, 219, 220, 238, 243, 250, 253, 272
Calcium chloride, 173, 243, 251
Calcium phosphate, 173
Calcium solution, 248
Calibration curves 36, 138–39, 143, 154
  bent, 173
  construction of, 144–47, 156, 158, 249, 251, 254, 257, 261, 267, 271
  interfering components, 149–51
  linearity of, 153
  slope of, 152, 165, 173
  translation of, 147
Capillaries, 55, 74, 79–81, 82, 86, 111, 122
CARIUS tube, 142
Cathodic sputtering, 118, 121
Cement production, 266
Ceramic materials, 260, 269
Chelate-forming reagents, 171
Chemical industry, 245–59
Chemiluminescence, 50–51, 201, 203
Chlorides, 22, 23
Chromates, 22
Chromatic aberration, 91
Chromium, 222, 233
Coagulation, 57, 168, 171
Cobalt, 234, 235
Coke furnace gas, 274
Collimator, 91
Collision dampening, 28, 32
Collisions, 50
Colloidal silicic acid, 171
Combustion, 37–40
  diffusion, 40
  kinetic, 40
  rate of, 41
Compressor, 66
Computer, 62
Concentration range, selection of, 138–40, 145, 161
Condition of electrical neutrality, 188
Continuous spectra, 16

Oscillator strength, 31, 204
Oxidants, 38, 40, 46, 52, 65
Oxides, formation of, 184–86
Oxygen, 67, 80

Palladium, 121, 235
Partial pressure of combustion gas components of acetylene–air flame, 183
Perchloric acid fluoride, 72
PERKIN–ELMER flame photometer, 85
  Model-146, 169
  Model-303, 123–24
pH value of solution, influence of, 184
Pharmaceutical industry, 274
Phosphates, 12, 23, 146–47
Phosphoric acid, 248, 250
Phosphorus, 236, 245
Phosphorus pentoxide, 247
Photocathode, 101, 102, 104
Photocells, 99
  barrier, 105–09
  fatigue in, 108, 132
  gas-filled, 102
  selenium, 112, 122, 128
  vacuum, 100–02
Photoelectric effect, barrier, 99
  external, 99, 100, 102
  internal, 99, 105
Photoelectric multipliers, 13, 98, 99, 103–05, 107, 113, 115, 122, 126, 164, 267
Photometer, 20
Photomultiplier. *See* Photoelectric multipliers
Photons, 25, 27
Photoresistors, 99
Phototubes, 12, 13
Physico-chemical processes, 62
PLANCK's constant, 25
PLANCK's law, 34
Plant physiology, 12
Plants, 269
Platinum-group metals, 235
POISEUILLE law, 55, 56
Portland cement, 269
Potassium, 117, 150, 151, 152, 189, 192, 193, 214, 222, 223, 239, 242, 245, 256, 260, 263, 269, 270
Potassium chloride, 241, 243, 246, 250, 260
Potassium hydroxide, 23
Potassium oxide, 253
Potassium solution, 253
Potentiometer, 135, 138
Pre-heating zone, 41
Pressure bottles, 65–66, 67, 70, 110, 127
Pressure nozzle, 78, 80, 122
Primary zone, 42
Prism, 16, 91–93, 113
Probability of absorption, 31
Probability of spontaneous emission, 31, 34
Probability value, 161
Propane, 69–70
Proportionality constant, 189
Pumps, dry type, 66

Qualitative analysis, 15, 21
Quantitative analysis, 11, 13, 21, 98
Quantometer, 13
Quantometric apparatus, 98
Quantum theory, 26
Quenching diameter, 73
Quenching effect, 22, 171, 176, 177–80

Radiant flux density, 30, 34, 35
Radiating layer thickness, 35
Radiation
  absorption, 25–26, 30–33
  background, corrections for, 195
  emission, 25–26, 33–37
  isolation, 88
  modulated, 123
  nature of, 25
  wavelength of, 15, 25, 134
Radiation detection, 12, 13, 33, 122

Radiation flux, 29, 30
Radiation intensity, 16, 21, 22, 45, 50, 62, 133, 143, 195, 203
Radiation interference, 88, 114, 157, 193–97, 204, 266
RAOULT law, 58
Rare earth elements, 227
RASCHIG rings, 69
RAYLEIGH criterion, 52
RAYLEIGH theory of disintegration of liquid column, 52
Reaction zone, 41, 42, 51
Reduction valve, 67, 127
Relative standard deviation, 160
Releasing effect, 180
Reliability interval, 161, 162
Residual emission, 175
Resonance dampening, 32
Resonance lines, 27, 30, 119, 126
Resonance profile, 27, 36
REYNOLDS number, 44
Rhodium, 235
Rocks, 259, 262
Rotating light chopper, 123
Rubidium, 21, 83, 117, 152, 155, 162, 189, 192, 193, 215, 262
Ruthenium, 235

Safety precautions, 71, 127
SAHA equation, 187, 191
Salt concentration, 139–40, 169
Salt solutions, 58
Sample inflow into flame, 51–63
Sample measurement, methods of, 147–48
Sample solution, dispersion of, 51
Samples
  decomposition, 142–43
  preparation, 142
  solid, introduction into flame, 82–83
Scandium, 227
SCHEIBE–LOMAKIN equation, 36
Secondary zone, 43
Sedimentation, 57, 168
Selenium, 14, 233
Selenium photocell, 112, 122, 128
Self-absorption, 34, 37, 138, 145, 172
Sensitivity, 162–66, 171, 203–04, 268
  definition, 165–66
Silicates, 83, 142–43, 259, 260, 261
Silicic acid, 259
Silver, 119, 225, 226
Silver iodide, 83
Slit width, 95, 134
Sodium, 46, 117, 141–42, 149–50, 151, 155–57, 173, 176, 177, 183, 189, 193, 195, 213, 221, 223, 239, 242, 260, 266, 274
Sodium chloride, 144, 241, 243, 250
Sodium hydroxide, 248
Sodium oxide, 253
Sodium solution, 253
Sodium sulphate, 141–42
Soil analysis, 269, 270, 271
Solid phase, interfering influences and their elimination, 171–81
Solutions
  pH value, 184
  preparation, 138–46
  surface tension, 57, 169
  viscosity, 168–69
Solvent molecules, diffusion coefficient, 59
Solvent partial pressure, 59
Solvent vapour pressure, 58–61
Solvents,
  amount dispersed greater than amount capable of evaporating in given volume, 59
  evaporation, 77
    in flame, 61
    in mist chamber, 58–61
  highly volatile, 60
  influence of, 48
  organic, 56, 61, 198–202
Specificity factor, 194–95
Spectral analysis, 98, 155
Spectral interval, 95
Spectral lamp, 117, 122

Conversion factors, 22
Copper, 21, 49, 119, 225, 226
Correction curves, method of, 151–52
Correction value, 152
Cosine law, 42
Cyanogen, 50, 71, 75

Decomposition of samples, 142–43
Degrees of freedom, number of, 159–60, 162
Diaphragm aperture, 133
Didymium filter, 194
Diffraction gratings, 93–94
Diffusion, loss by, 184
Diffusion coefficient of solvent molecules, 59
Diffusion combustion, 40
Diffusion flames, 40
Dilution, 238
Direct methods, 22, 23, 146
Discharge tubes, 117–23, 197
Dispersion, 51–53, 238
  basic principle, 78
  interfering influences, 168
Dispersion devices, 13, 52–53, 60–61, 74, 77–83, 111, 122, 140
  adjustment, 127
  angular-type, 80
  cleaning, 137
  concentric, 57, 80
  electrostatic, 81
  in parallel, 192
  location of, 84
  powered by external source, 81–82
  self-propelled, 78
  ultrasonic, 81
  with direct injection, 189
  with indirect injection, 192
Dispersion nozzles, crystallisation on, 60
Dissociation, 189
  degree of, 190, 204
  equilibria, 181, 187, 188, 191
  processes, 182
  rate of, 201
Dissociation constant, 182
Distillation, fractional, 62
Doppler effect, 27
Double-beam instrument, 13
Drop separator, 84, 111
Dynamic internal friction coefficient, 55

Electron, binding energy, 100
Electron liberation, 100
Electronegative elements, influence of, 192
Electronic excitation energy, 50
Electronics industry, 274
Elements
  absorption spectra, table of, 216
  emission spectra, 204
    table of, 206
  spectra, 213–36
Emission coefficient, 30, 104
Emission flame photometry, 16, 23, 202–03, 204
  classification of methods, 19
  interfering influences, 166
Emission spectra of elements, 204
  table of, 206
Energy levels, 26–27
Energy values, 28
Equilibrium concentration, change in, 183
Equilibrium constants, 181, 182, 190
Equilibrium shift in flame, 181–92
Evaporation, 61, 88, 178, 179, 238
  in mist chamber 58–61
  in vacuo, 61
  of solvent, 77
Excitation energy, 27, 50
Excitation mechanism in flame, 50
Excitation potential, 204
Explosion hazard, 66, 69, 112
External cone, 43
Extraction process, 250

"Fatigue" in photocell, 108–09, 132
Feed water, testing, 274
Fertilisers, 245, 269
Filter instruments, 132
Filters, 20, 21, 89–91
  coloured glass, 89, 122, 194
  didymium, 194
  glass, 13
  interference, 13, 89–91, 117, 122
  light, 88, 89
  properties of, 90
  transmittance peak, 89, 91
Flame conductivity, 191
Flame cone, size of, 43
Flame front, 41
Flame height, 43
Flame intensity, 172
Flame photometers, 64
  absorption, 117–26
  arrangement on bench, 127
  classification, 109
  compensation-type, 123–25, 134–35
  construction of, 125–26
  design of, 19
  double-beam, 116, 123
  emission, 109–17
  filter instruments, 132–33
  installation and adjustment, 127
  location, 126
  main parts, 65
  maintenance, 137–38
  measurement with, 132–37
  miscellaneous instruments, 114–17
  monochromator instruments, 133–37
  operating technique, 126–32
  overpressure, 66, 79, 129
  PERKIN–ELMER, 85
    Model-146, 169
    Model-303, 123–24
  principle of, 16–17
  schematic diagram, 17
  shut-down procedure, 133
  single-beam, 124, 125
  types, 109–26
  Unicam Model SP 900, 78, 85, 112–14
  Zeiss Model III, 78, 80, 83, 85, 88, 90, 110–12, 126, 127, 132, 137, 250
Flame photometry,
  advantages, 18
  applications. *See* Applications
  basic concepts, 18–21
  development, 202
  field of application, 24
  history, 11–15
  introduction to, 9–10
  methods of application of techniques, 21–24
  principles and characteristics, 15–18
  processes taking place in, 166
  significance of theoretical observations for practical analysis, 62
  techniques, 19
  theoretical considerations, 25–63
  use of term, 20
Flame profile, 184
Flame propagation, 41
Flame properties, 52
Flame spectrography, 19, 20
Flame spectrometry, 19, 20
Flame spectrophotometry, 19, 20
Flame spectroscopy, 19
Flame spectrum, 48–49
Flame structure, 40–45
Flame temperature, 12, 45–48, 51, 67, 70, 175, 176, 182, 192, 199, 200
Flames, 37–51
  acetylene, 190
  acetylene-air, 43, 47, 49, 112, 183, 189, 191, 200
  carbon–containing, 50
  colourless, 48
  cyanogen, 50
  cyanogen–oxygen, 50, 75
  diffusion, 40
  dinitrogen oxide–acetylene, 229
  evaporation in, 61

Flames *continued*
excitation mechanism in, 50–51
high-temperature, 71, 72, 203
hottest part of, 129
hydrogen, 51
igniting procedure, 129
laminar, 42, 43–45, 72, 73, 181
nitric oxide–acetylene, 72
nitrous oxide–acetylene, 71
oxygen–hydrogen, 50
poor in carbon, 51
sample inflow into, 51–63, 82–83
shift of equilibrium in, 181–93
stationary, 40, 41
town gas, 190
town gas–air, 191
turbulent, 44–45, 72, 75
weakly reducing, 186
Fluorides, 176
Fractional distillation, 62
Free atoms, 186, 191, 201, 204
Free radicals, 49
Fuel-to-oxidant ratio, 47
Fuels, 38–40, 46, 65, 67–72

Gallium, 229, 230
Galvanometer, 133, 144
screening, 128
Gaseous fuels, 38, 65
Gaussian distribution, 158
Gaussian profile, 27, 31, 36
Glass, 260, 269
Glycerine solution, 180
Glycerine suspension, 82
Gold, 225, 226
Grotrian diagram, 27
Group IV elements, 231
Group V elements, 231
Growth curve, 36

Half-width, 89
Hilger Uvispek spectrophotometer, 78, 125
Homogeneous radiant layer, 30
Hydrogen, 69
combustion, 39

Ignition temperature, 39, 41
Indirect methods, 22, 146
Indium, 83, 229, 230
Integral, 30
Integral absorption coefficient, 31
Integral emission coefficient, 30
Integral radiation flux density, 36
Intensity, radiation, 16, 21, 22
Interference factors, 22, 23
Interference filters, 13
Interfering components
and boiling point, 174–75
and emission intensity relationship, 173–74
elimination of influence of, 148–56
internal standard method, 155–56, 196
method of calibration curves, 149–51
method of correction curves, 151–52
method of model standards, 149
standard addition method, 152–55
Interfering influences, 166–202
elimination of
internal standard method, 169, 180
standard addition method, 170
of transport, and their elimination, 168–71
origins of, 166
shift of equilibrium in flame, 181–93
solid/gaseous phase conversion, 176
solid phase, 171–81
use of buffer solutions, 197
volatile matrix, 176–77
see also Organic solvents; Radiation interference
Internal standard method, 155–56, 169, 180, 196
Ion exchange, 178, 248
Ionic spectrum, 27

Ionisation, 172, 187, 189, 190, 192, 193
  degree of, 188, 189, 190
Ionisation constant, 187, 188, 193
Ionisation effect, 260
Ionisation energy, 27
Ionisation equilibrium, 187, 188, 191, 192, 193, 263
Ionisation potential 187
Iron, 121, 234

JOULE–THOMSON effect, 60

Kinetic combustion, 40
KIRCHHOFF law, 34
KIRCHHOFF theorem, 46
KJELDAHL method, 142

Laboratory record, 136
Lactate extracts, 270
LAMBERT–BEER law, 30
Laminar flow, 44, 55
Lamps. *See* Discharge tubes; Spectral lamps
LANGMUIR equation, 58, 59
Lanthanum, 180, 227
LAWRENCE–SMITH mixture, 83
Lead, 119, 231
Light alloys, 273
Limestone, 173, 176, 266
Limit of detection, 164, 165, 166, 204
Limit of determination, 165
Line intensity, 36
Line profile, 27, 28, 30
Line reversal method, 12, 45, 46
Line spectrum, 16, 17
Line spectrum sources, 117
Line width, 28, 31, 36
Lithium 46, 49, 83, 155, 156, 169, 183, 189, 193, 195, 214, 222, 260
LITTROW prism, 91

Magnesium, 119, 141, 177, 180, 184, 224, 264
Maintenance of flame photometer, 137–38
Manganese, 233, 234
Manometer, 111
Mercury, 117, 226, 227
MICHELSON law, 42
Microwave method, 191
Minerals, 259, 260, 262
Mist chamber, 72, 80, 83–88, 111, 122, 128, 137, 168, 181, 184, 198, 200
  evaporation in, 58–61
  optimum size of, 58
Mixing chambers, 111
Model standards, method of, 149
Molar ratio, 174, 175
Molecular spectrum, 29
Molybdenum, 186, 233, 234
Monochromator instruments, 133–37
Monochromators, 12, 13, 21, 28, 88, 91–97, 122, 126, 195
  resolution of, 94
  slit width, 95–97

Nickel, 234, 235
Niobium, 232
Nitrogen, 245
Noble metals, 119, 273
Non-metals, 236
Normaliser, 192, 193
Number of degrees of freedom, 159–60, 162

Occlusion, 175
Optica Milano Model CF 4 spectrophotometer, 114
Optical density, 32
Optical depth, 36
Order of spectrum, 94
Ores, 259
Organic complexes, 201
Organic reagents, 178–79, 186, 198–202
Organic solvents, 56, 61, 198–202
Organic substances, influence of, 186–87

Oscillator strength, 31, 204
Oxidants, 38, 40, 46, 52, 65
Oxides, formation of, 184–86
Oxygen, 67, 80

Palladium, 121, 235
Partial pressure of combustion gas components of acetylene–air flame, 183
Perchloric acid fluoride, 72
PERKIN–ELMER flame photometer, 85
  Model-146, 169
  Model-303, 123–24
pH value of solution, influence of, 184
Pharmaceutical industry, 274
Phosphates, 12, 23, 146–47
Phosphoric acid, 248, 250
Phosphorus, 236, 245
Phosphorus pentoxide, 247
Photocathode, 101, 102, 104
Photocells, 99
  barrier, 105–09
  fatigue in, 108, 132
  gas-filled, 102
  selenium, 112, 122, 128
  vacuum, 100–02
Photoelectric effect, barrier, 99
  external, 99, 100, 102
  internal, 99, 105
Photoelectric multipliers, 13, 98, 99, 103–05, 107, 113, 115, 122, 126, 164, 267
Photometer, 20
Photomultiplier. *See* Photoelectric multipliers
Photons, 25, 27
Photoresistors, 99
Phototubes, 12, 13
Physico-chemical processes, 62
PLANCK's constant, 25
PLANCK's law, 34
Plant physiology, 12
Plants, 269
Platinum-group metals, 235
POISEUILLE law, 55, 56
Portland cement, 269
Potassium, 117, 150, 151, 152, 189, 192, 193, 214, 222, 223, 239, 242, 245, 256, 260, 263, 269, 270
Potassium chloride, 241, 243, 246, 250, 260
Potassium hydroxide, 23
Potassium oxide, 253
Potassium solution, 253
Potentiometer, 135, 138
Pre-heating zone, 41
Pressure bottles, 65–66, 67, 70, 110, 127
Pressure nozzle, 78, 80, 122
Primary zone, 42
Prism, 16, 91–93, 113
Probability of absorption, 31
Probability of spontaneous emission, 31, 34
Probability value, 161
Propane, 69–70
Proportionality constant, 189
Pumps, dry type, 66

Qualitative analysis, 15, 21
Quantitative analysis, 11, 13, 21, 98
Quantometer, 13
Quantometric apparatus, 98
Quantum theory, 26
Quenching diameter, 73
Quenching effect, 22, 171, 176, 177–80

Radiant flux density, 30, 34, 35
Radiating layer thickness, 35
Radiation
  absorption, 25–26, 30–33
  background, corrections for, 195
  emission, 25–26, 33–37
  isolation, 88
  modulated, 123
  nature of, 25
  wavelength of, 15, 25, 134
Radiation detection, 12, 13, 33, 122

Radiation flux, 29, 30
Radiation intensity, 16, 21, 22, 45, 50, 62, 133, 143, 195, 203
Radiation interference, 88, 114, 157, 193–97, 204, 266
RAOULT law, 58
Rare earth elements, 227
RASCHIG rings, 69
RAYLEIGH criterion, 52
RAYLEIGH theory of disintegration of liquid column, 52
Reaction zone, 41, 42, 51
Reduction valve, 67, 127
Relative standard deviation, 160
Releasing effect, 180
Reliability interval, 161, 162
Residual emission, 175
Resonance dampening, 32
Resonance lines, 27, 30, 119, 126
Resonance profile, 27, 36
REYNOLDS number, 44
Rhodium, 235
Rocks, 259, 262
Rotating light chopper, 123
Rubidium, 21, 83, 117, 152, 155, 162, 189, 192, 193, 215, 262
Ruthenium, 235

Safety precautions, 71, 127
SAHA equation, 187, 191
Salt concentration, 139–40, 169
Salt solutions, 58
Sample inflow into flame, 51–63
Sample measurement, methods of, 147–48
Sample solution, dispersion of, 51
Samples
  decomposition, 142–43
  preparation, 142
  solid, introduction into flame, 82–83
Scandium, 227
SCHEIBE–LOMAKIN equation, 36
Secondary zone, 43
Sedimentation, 57, 168
Selenium, 14, 233
Selenium photocell, 112, 122, 128
Self-absorption, 34, 37, 138, 145, 172
Sensitivity, 162–66, 171, 203–04, 268
  definition, 165–66
Silicates, 83, 142–43, 259, 260, 261
Silicic acid, 259
Silver, 119, 225, 226
Silver iodide, 83
Slit width, 95, 134
Sodium, 46, 117, 141–42, 149–50, 151, 155–57, 173, 176, 177, 183, 189, 193, 195, 213, 221, 223, 239, 242, 260, 266, 274
Sodium chloride, 144, 241, 243, 250
Sodium hydroxide, 248
Sodium oxide, 253
Sodium solution, 253
Sodium sulphate, 141–42
Soil analysis, 269, 270, 271
Solid phase, interfering influences and their elimination, 171–81
Solutions
  pH value, 184
  preparation, 138–46
  surface tension, 57, 169
  viscosity, 168–69
Solvent molecules, diffusion coefficient, 59
Solvent partial pressure, 59
Solvent vapour pressure, 58–61
Solvents,
  amount dispersed greater than amount capable of evaporating in given volume, 59
  evaporation, 77
    in flame, 61
    in mist chamber, 58 – 61
  highly volatile, 60
  influence of, 48
  organic, 56, 61, 198–202
Specificity factor, 194–95
Spectral analysis, 98, 155
Spectral interval, 95
Spectral lamp, 117, 122

Spectral line, 27
 and bands, table of, 275
Spectral slit width, 95, 134
Spectrograph, 12, 19, 20, 97–98
Spectrometer, 20
Spectrophotometers, 122, 124
 BECKMAN, 114
 Hilger Uvispek, 78, 125
 maintenance, 138
 Optica Milano Model CF-4, 114
Spectroscope, 16
Spectrum, 15, 19, 21, 23, 26
 atomic, 26, 27
 band, 16
 continuous, 16
 flame, 48–49
 ionic, 27
 line, 16, 17
 molecular, 29
 of elements, 213–36
 order of, 94
Standard addition method, 152–55, 170
Standard deviation, 159–62, 164, 165
Standard solutions, 146, 148, 149, 150, 156
 composition, 262
 preparation, 140–42, 240, 262, 279
 storage, 141
Statistical security, 161, 162, 164
Steel, 273
Stoichiometric relations, 22
STOKES' law, 57
Strontium, 153, 155, 176, 177, 180, 184, 195, 219, 222, 260
Sulphates, 12, 22, 178, 180
Sulphide ores, 143
Sulphur, 236
Surface tension, 57, 169

Tantalum, 232
Technical trisodiumphosphate, 247
Tellurium, 233
Thallium, 83, 117, 229, 230
Thermistor, 117
Tin, 186, 201, 231
Titanium, 231
Town gas, 70–71, 111, 131
Townsend discharge, 121
Transition elements, 197
Transmittance coefficient, 96
Transport effects, 191, 198
 interfering influences, 168–71
Tungsten, 233
Tungsten chlorides, 255
Tungstic acid, 253
Turbulent flow, 44

Ultrasonic effects, 81
Unicam Model SP 900 flame photometer, 78, 85, 112–14
Urine, 272

Vacuum filling apparatus, 119–21
Vacuum photocells, 100–02
Valve, reduction, 67, 127
Vanadates, 178
Vanadium, 178, 232
Vanadium contact masses, 256
Variation coefficient, 160
Variation range, 160
Ventilation, 127
Viscosity of solution, 168–69
Voigt profile, 36
Volatile matrix, 176–77
Voltage divider, 104
Volumetric determinations, 23

Water analysis, 237–45
Water softener, 247
Wavelength, 15, 25, 234
Wavelength range, 88, 102
WEBER number, 54

Yttrium, 227

Zeiss Model III flame photometer, 78, 80, 83, 85, 88, 90, 110–12, 126, 127, 132, 137, 250
Zinc, 117, 119, 226, 227